Prophets of the Posthuman

CHRISTINA
BIEBER LAKE

Prophets

of the

Posthuman

American Fiction, Biotechnology,
and the Ethics of Personhood

University of Notre Dame Press
Notre Dame, Indiana

Library of Congress Cataloging-in-Publication Data

Lake, Christina Bieber.
Prophets of the posthuman : american fiction, biotechnology, and the
ethics of personhood / Christina Bieber Lake.
pages   cm
Includes bibliographical references and index.
ISBN 978-0-268-02236-5 (pbk. : alk. paper) — ISBN 0-268-02236-4
(pbk. : alk. paper)
1. American fiction—20th century—History and criticism. 2. Ethics in
literature. 3. Bioethics in literature. 4. Human beings in literature
5. Literature and technology—United States—History—20th century.
I. Title.
PS374.E86L35   2013
810.9'353—dc23
                                        2013022548

To VTS,

of course

Contents

Acknowledgments

I cannot imagine completing this project without my dear friends who so willingly read the manuscript at various points: Tiffany Kriner, Nicole Mazzarella, and Beth Felker Jones. You women are irreplaceable. I would also like especially to thank Alan Jacobs for his advice, encouragement, and assistance, and David Wright for helping me, years ago, to see the next step.

I am grateful to C. Ben Mitchell, Nigel Cameron, Jennifer Lahl, Brent Waters, Jill Baumgaertner, Tim McIntosh, Avis Hewitt, John Sykes, and Amy Laura Hall for encouraging my nascent idea that a literary scholar could contribute to the conversation in bioethics. I also appreciate the Tumblers: the faculty members of the Wheaton College 2009–10 advanced Faith and Learning seminar who were excellent sounding boards along the way. Stephen Little and all the folks at the University of Notre Dame Press have been a delight to work with; thank you for supporting this project.

So many of my students have helped that I am afraid of failing to mention them all. First, I would like to thank the seniors who attended my seminars on literature and the posthuman whose contributions to class and enthusiasm helped me immeasurably. Special thanks goes to Aubrey Penney, who spent hours editing the text and endnotes, and Elise Bremer, Rachael Shaffner, Rachel Maczuzak, Will Hierholzer, Heather Fredricks, Alec Geno, Abby Long, and Tara Newby for similar help.

Alan Savage and Sandy Oyler: thanks for being a part of our family and for enduring my current scholarly "phase." Last and never least, my dear husband, Steve: thank you for believing in this project and in me.

Preface

Only Evolve! Bioethics and the Need for Narrative

> *The primary political and philosophical issue of the next century will be the definition of who we are.*
>
> —Ray Kurzweil in 1999

Ray Kurzweil is afraid to die.

Multimillionaire inventor of the first reading machine for the blind, Kurzweil is best known for his predictions about the future that culminated in his 2009 book, *The Singularity Is Near.*[1] Kurzweil predicts that by 2045 machines will exceed human intelligence and the posthuman era will begin, eventuating in solutions to all of our most pressing problems, including death. In the opening voice-over of the recent documentary *Transcendent Man,* Kurzweil speaks slowly and deliberately, with haunting strains of the music of Philip Glass in the background:

> I do have a recurring dream. It has to do with exploring this endless succession of rooms that are empty, and going from one to the next, and feeling hopelessly abandoned and lonely and unable to find anyone else. That's a pretty good description of death. Death

is supposed to be a finality, but it's actually a loss of everyone you care about. I do have fantasies sometimes about dying. About what people must feel like when they're dying, or of what I would feel like if I were dying. And it's such a profoundly sad, lonely feeling . . . that I really can't bear it. And so I go back to thinking about how I'm not gonna die.[2]

There is nothing new, of course, about Kurzweil's fears or hopes. The inevitability of death has always shaped human psychology, philosophy, religion, and the arts. What is relatively new here is the specific content of Kurzweil's optimism: he believes that his life on Earth will literally not end. In his lifetime, humanity will evolve to overcome death by learning how to repair diseased and aging cells, and eventually how to download minds into computers.[3] Kurzweil's personal desires have become a part of his prophetic narrative: by way of the exponentially increasing power of science applied through technology, humans will return to the garden of Eden, with not only a new Eden but a new Adam and a new Eve to inhabit it. And Kurzweil is far from alone in this ultimate prediction. When Lee Silver, a Princeton biologist, wrote *Remaking Eden: How Genetic Engineering and Cloning Will Transform the American Family,* this is what he meant: that our scientific knowledge and technical skill will ultimately give us complete control over our own evolutionary future. "We, as human beings, have tamed the fire of life," Silver writes, describing this future world. "And in so doing, we have gained the power to control the destiny of our species."[4]

Whether the ability to control the destiny of the human species will turn out to be a good thing remains to be seen. Either way, to define transcendence as the inevitable outcome of technologically driven human evolution represents not only a phenomenon unique to the twentieth and twenty-first centuries[5] but also a rejection of thousands of years of philosophical and theological thinking about what constitutes the highest and best life available to human beings.[6] While it is tempting to think of Kurzweil and Silver as outliers, their thinking is merely a logical extension of the increasing confidence that late modern people have placed in finding technological solutions to problems. This belief could be summed up by the mantra "Only evolve!" This kind of evolu-

tion, it must be noted, is not Darwinian evolution; it assumes that
Mother Nature has been fickle and random, and that we can and should
do much, much better.[7] Variations of this mantra can be seen in best-
selling books by Steven Pinker, Daniel Dennett, Lee Silver, Simon
Young, Rodney Brooks, and many others. What they all share is the
belief that we inherently know what the good life is (to be free from suffering, disease, death, and other difficulties) and that it is something
that we can and must *make,* not learn.[8] Technoscience—scientific knowledge applied through technology—is the way to make that life.

As if this change were not profound enough, the "Only evolve!"
mandate resists any challenges to its fundamental definition of the good
life. But that doesn't phase Kurzweil or any of these thinkers, for as a
mandate built on a scientific naturalist conception of human life, it has
no mechanism for self-questioning. Eric Cohen puts it very simply:
"science is a means to many ends without wisdom about which ends
are most worthy."[9] Consumer culture is left to itself to define the ends
in the form of products and services that affluent Americans stand by
ready to purchase.[10] Thus, as Brent Waters has argued, the best way to
characterize the goal of late modern technology is not by modern conceptions of progress but by a desire to transcend limitations simply
because they are limitations.[11] Kurzweil insists that what "represents the
cutting edge of the evolutionary condition" is simply "to seek greater
horizons and to always want to transcend whatever our limitations are
at the time."[12] What Kurzweil names as a limitation or how he plans to
transcend it is not the issue. It does not matter what we are evolving
into, only that we evolve. What matters most about our destiny is
simply the fact that we get to choose it.

As I will develop in the introduction to this book, confidence in the
mandate to "only evolve" has serious implications for ethics. It changes
both the urgency and the shape of the ancient philosophical question,
how should we then live? Questions about how to gain immortality or
psychological equilibrium or enhanced cognitive function replace inquiry into the desirability of attaining those things. In a 2006 debate
with Ray Kurzweil, Baroness Susan Greenfield, a neuroscientist, pressed
him on the issue of immortality. She asked Kurzweil what exactly we
think we would do with eternal life and whether having longer lives

would necessarily make people happier. He never really answered the question.[13]

America faces rapidly accelerating technological change in the area of human enhancement, and that change is led by Kurzweil-type prophets who do not converse with, consult, or question leaders from other disciplines about the desirability of their vision.[14] This is precisely the outcome that Hannah Arendt insisted we must do everything we can to avoid:

> This future man, whom the scientists tell us they will produce in no more than a hundred years, seems to be possessed by a rebellion against human existence as it has been given, a free gift from nowhere (secularly speaking), which he wishes to exchange, as it were, for something he has made himself. There is no reason to doubt our abilities to accomplish such an exchange, just as there is no reason to doubt our present ability to destroy all organic life on earth. The question is only whether we wish to use our new scientific and technical knowledge in this direction, and this question cannot be decided by scientific means; it is a political question of the first order and therefore can hardly be left to the decision of professional scientists or professional politicians.[15]

These decisions have, in fact, been left to scientists and politicians. They are not being handled publicly in the manner Arendt advocates. Instead, the humanistic disciplines that would be most interested in the philosophical question behind these decisions—namely, how should we then live?—have instead doubted the value of addressing that question in the academy. A variety of thinkers have sketched out this so-called crisis in the humanities, which is often described as a forceful dismissal of such bluntly ethical questions.[16] This crisis has been most pronounced in literary studies, which, toward the end of the twentieth century, had all but abdicated thinking of itself as a generally humanistic discipline that contributes scholarship to larger political conversations in favor of thinking of itself as a guild of specialists who contribute scholarship accessible only to a narrow band of academics.[17]

In their efforts to historicize the crisis, Brian Stock and Martha Nussbaum argue that the way individuals read (especially the way we read fiction) has changed significantly because of this arrangement. Stock compares the goals of modern readers to those of ancient readers like Seneca and Augustine, who believed in the reader's "ethical responsibility for postreading experience."[18] For them, the hard work began after the text was read, in applying its vision to gain self-awareness and to grow in the virtues. From that position individuals could make meaningful contributions to the *polis*. The modern reader, on the contrary, locates the work to "some form of interpretation, that is, to expounding, clarifying, or explaining the text."[19] The ancient question of the good life is left behind, and "interpretation has become the only widespread postreading activity."[20] As Martha Nussbaum explains, the disciplines of both philosophy and English converged to exclude ethics from inquiry. Philosophers were reluctant to ask the ancient question, how should we then live?, and English professors were even more so. This reluctance was due partly to the discipline's efforts to keep ethicists from simply raiding texts as if they were nothing but pantries full of bite-sized morals. Some of the reluctance came from the effort to justify English as its own discipline, and some came from the theoretical skepticism of poststructuralist theory. Regardless of the motives, mid- to late-twentieth-century literary critics often assumed, as Nussbaum explains, that any critical work interested in the reader's practical needs "must be hopelessly naïve, reactionary, and insensitive to the complexities of literary form and intertextual referentiality."[21]

One of the results of the exclusion of the ethical question from the academy is the production of a kind of rift in reading practices among contemporary Americans, where professional readers—scholars of literature—have little influence over how, why, or what the average American reads. John Guillory explains the loss as one of the "intermediate practices of reading," which lie "between the poles of entertainment on the one side, and vigilant professionalism on the other."[22] The two types of reading are so disconnected that they barely resemble one another. So Guillory concludes that the only way to understand reading as a social practice is by returning, in some way, to the ethical.[23]

Although there has been a recognized "turn toward the ethical" already (which I will describe below), the continued misrecognition of reading as an inherently ethical activity has impoverished public debate on questions that reach beyond the traditional domains of literary study. By isolating fiction between the two poles of reading professionally and reading for entertainment, fiction's potential contribution to the larger ethical debates is marginalized. This has happened at exactly the time when, because of the rapidity of change in the biotechnological revolution, its need may be most urgent.

For example, one result of the failure to recognize the ethical importance of reading fiction is that the human enhancement conversation has been sequestered within the discipline of bioethics. Not surprisingly, bioethicists are trained to discuss very specific scenarios, and not how to read fiction in order to draw its much-needed insight into the debates. Indeed, a quick look at the ethical debates surrounding human enhancement technologies reveals that fiction is not a wise old sage that scholars turn to for discerning moral inquiry. Fiction is treated less like a sage and more like a slave; it is used to churn the wheel of argument or is simply ignored as irrelevant. For example, even Leon Kass, who is one of the better readers of fiction among bioethicists, is not always consistent in treating novels like Aldous Huxley's *Brave New World* as the complex and flawed texts that they are. This encourages scientists like Stephen Pinker to debase fiction and the role it can play by merely dismissing it, belittling Kass for treating "fiction as fact."[24]

In spite of these problems, in recent years some very good writers have paved the way for the return of narrative to ethical questions.[25] Geoffrey Galt Harpham, writing in a special issue of *New Literary History,* issued a plea for literary scholars to be willing to write for a larger audience, to contribute "in some identifiable way to a purpose beyond that of the accumulation of knowledge for its own sake."[26] Specifically, scholars can demonstrate how textual studies can offer what Carla Hesse explains is vital to ethical reflection: "deep investigation, concentration, reflection, and contemplation."[27] Wayne Booth argues that the very ubiquity of stories means they are already a part of our ethical reflection—the work they do just needs to be highlighted. Whether we acknowledge it or not, we are all affected by stories, and everyone feels the ethical effects of engaging with them: "No human being, literate or not, escapes the effects of stories, because everyone tells them and listens to them."[28] While critics may think that the best thing to do is to put themselves above identifying with the texts, argues Booth, actual readers do not so purge themselves of these responses. That is reason enough to take them seriously. And all of these scholars insist that to take stories seriously as ethical barometers is not to ignore the question of form and other questions distinctive to the discipline of English—far from it. As Nussbaum reminds us, even Plato, that most famous banisher of the poets, knew that "to choose a style is to tell a story about the soul."[29] In other words, not only is the story about who we are and what we want to become, but it also enacts those options in a way that nothing else can. Narrative must be treated in all of its complexity.

One of my goals in writing *Prophets of the Posthuman* is to demonstrate that ethical debates—if they are to be meaningful at all—require deep, nuanced, and ongoing reflection on narrative.[30] Narrative does not visit ethical questions abstractly; it lives them, because it lives in the realm of ethos, of persons as persons engaged with one another. Persons influence each other for good or ill; persons love one another well or poorly. Persons are the ones in pursuit of the good life; persons need to find out why they are alive at all. The fact that all these things are true in part explains why the novel was born and why it still flourishes today.[31]

In short, precisely because it is the question, what is the good life for persons?, that is at issue, *Prophets of the Posthuman* insists that the humanities, and especially literary study, can no longer take a backseat when it comes to bioethics. As Harpham discerns, there is a new urgency for conversation between humanistic and nonhumanistic disciplines as "they confront not only such new subjects as genetic engineering, environmental trauma, and the cognitive capacities of animals or machines, but also, and most intriguingly, such traditional subjects as the nature of language and the distinctive features of a specifically human being." Because these issues concern fundamental questions of humanity, the humanities, whose realm has always been the larger world of meaning and values, must now reposition itself as the "natural sponsor of the debates and controversies that swirl around such issues."[32]

To become such a sponsor, literary studies must not be afraid to trace how and why contemporary novels in particular have directly and indirectly intersected with the core issues of posthuman enhancement technologies. And one of the first things that needs to be addressed is that while some literary scholars may have been content to avoid moral and ethical questions, novelists themselves thrive in them, even those who claim, like Oscar Wilde, to avoid them as a matter of principle.[33] Though they might not put it precisely this way, most writers would resonate with George Saunders's belief that "all good fiction is moral, in that it is imbued with the world, and powered by our real concerns: love, death, how-should-I-live."[34] Even writers who find this statement repugnant would agree that they are interested in imparting their vision—whatever it might be—to their readers. One could even say that vision is their trade, the goal of their work, their raison d'être. In the well-known words of Joseph Conrad, the writer's task is "to make you hear, to make you feel—it is, before all, to make you see. That— and no more, and it is everything."[35]

This concern with vision means that although not a few of them might shun the label, many fiction writers are also prophets. Like all prophets, they try to speak in venues where the largest number of people might be able to see and hear, to change their vision of the world, even to change their vision of change. The venue they have chosen is not the pulpit or the classroom, but something they see as even more central and vital to the formation of vision: fiction. In fiction, they flesh out the worlds we think we want; they imagine the outcome of our deepest dreams; they challenge the desires that fuel our decisions. Their work insists that stories are not a luxury. We live them; we need them. We need them because we live them. Novelists have recognized the pressure that technology puts on what is good for a human being: what this life *means* and what it is *for*.[36]

The writers I engage with in this book are not only concerned with this question but concerned that their readers be sufficiently interested in it. For all but perhaps the most crass of metafiction writers, literary artists maintain a hope that someone in their audience will see things the way they see them. These writers want individuals to reconsider how they see themselves, others, and life itself. George Saunders is just such a writer. In an interview, he remarked that reading Toni Morrison's *The Bluest Eye* helped him to recover his sensibility that fiction can be about this kind of transformation: "The idea is that you go into this room called the short story and come out different. I don't really care what's in the room, as long as when you come out, you're 6% more aware, more happy to be alive, more appreciative, more curious, instead of closed down."[37]

Prophets of the Posthuman is fundamentally a plea for us to venture into the house of fiction and stay awhile. Its rooms contain something that no one interested in ethics can do without: the real people who will be affected by our decisions, good and bad. The very best fiction writers stand apart from their merely popular counterparts because they do not allow us to assume that any vision is correct because it feels good, sounds right, or sells well. They dare to ask what life is for and whether we are aiming for what would really be best for us. These writers are the ones brave enough to venture under the foundation of our posthuman dreams and find out what is underneath them, what may be rotting, or what we have tried to stuff down there in order to forget.

This book unveils their vision.

Introduction

Learning to Love in a Posthuman World

> *Let us treat the men and women well: treat them as if they were
> real: perhaps they are.*
>
> —Ralph Waldo Emerson, "Experience"

In 1836 Ralph Waldo Emerson published an essay called "Nature" that is not about nature at all. It is about how the triumphant American self should interact with the received world, with whatever it perceives as limitations, and that is by refusing those limitations. Americans must believe that "Nature is not fixed but fluid. Spirit alters, moulds, makes it." Americans must build our own worlds, must shape our lives to fit our greatest aspirations. If we do that, promises Emerson, the physical world will follow suit. "A correspondent revolution in things will attend the influx of the spirit. So fast will disagreeable appearances, swine, spiders, snakes, pests, mad-houses, prisons, enemies, vanish; they are temporary and shall be no more seen."[1]

The language here is exaggerated, of course, and easy to mock. But it achieves its hortatory purpose. Americans have always been idealistic, and Emerson knew that to attain a better life one must be able to envision it. A lot of good has come from individuals refusing to accept disagreeable realities; some of the greatest social reformers lived in Emerson's time. After all, everyone can agree that the fewer snakes, spiders, and enemies we have to deal with, the better.

But contemporary Americans also live in a time separated from Emerson's by two revolutions: the industrial and the technological. Something pernicious happened to Emerson's idealism as it lingered into the late twentieth and early twenty-first centuries. It narrowed, coupled with technological optimism, and morphed into something quite different. Contemporary biotechnological advances make Emerson's "revolution in things" more *literally* possible than ever. Through what some are calling "superbiology," techno-utopians promise an actual end to anything that gets in our way: prisons will disappear once we find the genetic markers for criminal behavior, pests will be under control once we find a way to safely kill them, our cantankerous spouses will love us better once we perfect mood-altering drugs, and even death will no longer be a limit once we find the biological keys to aging. Idealism has now met the age of DNA. When they merge, these new idealists promise us, we will become completely posthuman and completely free.

To choose just one example from among many, in 2006 Simon Young published a book entitled *Designer Evolution: A Transhumanist Manifesto*. Transhumanists are a loosely organized group of people who believe that science and technology can and should be used to overcome all human limitations. Although his prose is much less elegant, Young's ideas sound very similar to Emerson's: "Why must we age and die? Why must our brains and bodies be so fragile, subject to inevitable decay— programmed for self-destruction? I believe in seeking to overcome the mental and physical limitations that restrict our freedom. Science offers the only serious possibility of succeeding. Therefore, I believe in science."[2] The most significant difference here, of course, is that whereas Emerson gave the human imagination the power to transcend all earthly limitations, Young gives that power to applied science alone. He even coins the word "Neuromanticism" to describe how his is the same goal with different means. Neuromanticism, with its focus on altering the brain and securing genetic perfection, seems in many ways to be the ultimate answer to the problems Emerson outlined in his late essay "Fate," in which he bemoans how quickly we run into limitations that are truly immovable. Emerson ponders how we want to reform humans, but when we try to educate some children, "we can make nothing of them. We decide that they are not of good stock. We must begin our reform earlier still,—at generation: that is to say, there is Fate, or laws of the world."[3] Emerson saw these "laws of the world" as a more or less impassable barrier, and he died in despair. But for Young, conquering the laws of the world is simply the next step. Young believes that technology will allow us to re-create humanity, to fix "butterfingered" Nature's mistakes once and for all: "Where *Homo sapiens* was the slave of his selfish genes, *Homo cyberneticus* will be the steersman of his own destiny."[4] Technology will be the ship by which *Homo cyberneticus* will steer himself into ultimate freedom.

So what's the problem? The destiny described by the proponents of self-evolution is not just the destiny of the self. Because these changes involve the literal recreation of the human body, technological evolution is necessarily a corporate destiny involving millions of other people, especially children. This is fundamentally an ethical issue. But Young, Emerson, and other American idealists share a rhetorical strategy that assumes mutual agreement on goals without asking about the actual lives of the people they set out to fix. Furthermore, and perhaps even more importantly, both Emerson and Young assume it is right to try to achieve the desire for transcendence and control in the future, but rarely ask how these desires influence the way we treat others now. Emerson is not known for his ethical writings, and when he does venture to talk about the claim the other might have on the self, he comes out with weird statements like the one in the epigraph: "Let us treat the men and women well: treat them as if they were real: perhaps they are." The statement reveals the ethical conundrum at the core of most theories of personal transcendence, whether that transcendence is through imagination or technology. It is as if Emerson is saying, "what matters most is that I go forward with my freedom to change the world by starting with myself. Perhaps others have real claims on me, perhaps not, but let us go ahead and assume them to be real, and treat them well." But in a world where biotechnology is giving us startling new powers to literally and permanently shape the next generation, what does it mean to treat people well *now?*[5] If I value my personal freedom highest, what happens to my neighbor? Can we simply lay aside, as Emerson does here, the question of how real the other is? If we do, what in this philosophy will teach me what it means to treat someone well—to want for them the good life?

Put most simply, ethics is concerned with decisions that affect the self and the other as they are in relation with one another. And theoretical inquiry, at least since Alisdair MacIntyre's *After Virtue,* reveals the substantial difficulties that starting with the autonomous, enlightened self (such as Emerson extolled) causes for ethics. Because ethics involves the determination not only of what is good for the self but also what is good for the other, it requires an acknowledgement of persons as bearers of particular roles in a society—as neighbors, parents, friends, members of gilds, citizens of nations—not just as individuals. These roles constitute my "moral starting point" and require responsibility; they are not "merely contingent social features of my existence."[6]

MacIntyre is hardly alone in his concern; philosophers who reject personal autonomy as a starting point in order to instead begin with this category of personal responsibility to the other (especially as seen in narrative) come from a variety of different traditions that certainly do not agree with one another on all the issues. To start with the notion of our responsibility to the other, for example, compels neither the continental philosopher Immanuel Levinas nor the Catholic philosopher Robert Spaemann to agree with MacIntyre's larger project to return to an Aristotelian notion of the virtues. But in spite of vast areas of disagreement, the philosophers I draw on in this book agree on one salient issue: that ethics that begins with one's personal freedom, instead of one's personal responsibility to the other, is doomed to fail.[7] Put another way, as Jacques Maritain and many others have argued, ethics requires a definition of "persons" that insists on persons as wholes that exist within, and depend on, society as a whole, not merely as materially independent "individuals" whose freedom is found primarily by self-determination.[8]

A look into current bioethical inquiry will reveal how difficult it is to learn how to be ethically responsible to another person when the desire to be free from limitations is one's starting point. In his book

The Future of Human Nature, Jürgen Habermas focuses on two of the most challenging issues in bioethics: the various uses of PGD (pre-implantation genetic diagnosis) and the promise of germline genetic engineering. Germline genetic engineering—the selecting and passing on of certain genetic traits to one's offspring—is not yet possible, but PGD, the creating and then selecting from certain embryos through IVF (in vitro fertilization), is. PGD has been used, for example, by parents who carry the gene for Huntington's disease. While few will argue with this particular choice, that fact alone is not what concerns Habermas. What concerns him is that in PGD, parents will choose life for certain embryos and death for others based solely on their views of which child is most desirable and will have the best chance at happiness, as the parents define it. That the choice is the parents' alone is the rub. Habermas argues that while philosophy had formerly been concerned with the good life as a question of communal justice, today it is left up to individuals. In short, ethics has supplanted moral philosophy, and ethics is "I" oriented. The new technologies make this transition blazingly clear, and we are now seeing "the instrumentalization of conditionally created human life according to the preferences and value orientations of third parties."[9] In other words, individuals in our society not only decide what the good life is for themselves but make permanent, lasting decisions about the good life for *others.* We have moved from the realm of the other as given to the realm of the other as made. As a result, the question of what it means to treat others well, to treat them as if they were "real," has an unprecedented new salience.

Habermas's answer to this shift is provocative as far as it goes. A good Kantian ethicist, Habermas argues that since the new technology often involves making decisions for others, we must especially guard against using others as objects. Since we are in a "postmetaphysical" situation that can no longer see God as the ultimate other (in other words, there is no widespread acceptance of the idea of being made in God's image), we must work together through language to agree on the way we should treat others. One of the ways that Habermas thinks this can happen is by stressing that decisions be guided by the principle of thinking of the other as a physician would think of a person, which is as someone she will again encounter face to face, and not as a technician would, which is "as an object which is manufactured or repaired or channeled into a desired direction."[10] Then Habermas explains that "what solely matters here is not the ontological status of the embryo, but the clinical attitude of the first person toward another person—however virtual—who, some time in the future, may encounter him in the role of a second person."[11] Thus, it is not genetic engineering that is the issue, but the attitude with which the interventions are carried out. In gene therapy, for instance, we have the correct attitude because we are thinking about the person the embryo will be.[12] How we see the other person is the main thing that determines how we ascertain our duty to them.

Although Habermas makes the vital (if surprisingly rare) point that the ethical validity of one's actions toward another is determined primarily by one's attitude toward the other, he still makes this argument starting from a defense of individual autonomy. That is to say, his argument is centered around the conviction that since the other is or will be a person, she has the same rights to freedom that I do, and so I should afford her those rights by not intervening too much. But a close look at his text reveals that caring about the other's autonomy before caring about one's general responsibility to the other has left Habermas with no answer to some crucial questions. If we put aside the question of whether the other is real, where does the attitude of the medical professional or the parent come from? What determines it, shapes it? What will *cause* someone to think of the other as a person rather than an object? Laurie Zoloth argues quite convincingly that if bioethical theory rests first and foremost on the principle of the autonomous self, then selfhood starts with freedom.[13] And if it does, then it quickly follows that the self is a free consumer, potentially reducing the world, including other people, to "a series of things one can possess." She then explains that autonomous desire contains a "hidden request" for the other to act on one's own behalf. In other words, if the self begins with freedom, it is expressed chiefly with power. When considered as our primary ideal, autonomy is more likely to encourage us to view the other instrumentally, as a thing to be used, rather than as a person who deserves to be cared for or even to be loved. Zoloth then provocatively asks: With that priority, "can any desire be achieved without the use of the other in the service of the self?"[14]

This is a terrific question. In an age in which many bizarrely individuated desires are more attainable than ever before, the fact that few people ask it is chilling. The point of this book is to argue that we must ask where our attitudes toward the other (and toward our own bodies) come from. We especially must consider how having personal autonomy as one's highest ideal might militate against what could be called an ethics of hospitality. We cannot just say we will treat men and women well because they *may* be real; we have to ponder what exactly causes us to decide whether they are real or not when they come knocking on our door. Though Miroslav Volf's *Exclusion and Embrace* deals with ethnic and cultural conflict, his point about ethics can be applied here. Modern society too quickly defaults to questions of "social arrangement" instead of thinking about the kind of "social agents" it fosters. Without collapsing into promoting vapid personal piety, Volf argues convincingly that the question is not what kind of society we need to have, but what "*kind of selves we need to be* in order to live in harmony with others."[15] What is urgently needed is "reflection on the character of social agents and of their mutual engagement."

If Volf is correct—and I believe that he is—then narrative is the most logical place to do this work. Narrative, as many scholars have shown, is of special interest to the kind of virtue-oriented ethics that I am arguing for.[16] Narrative suggests the kinds of selves we can be; it gives outcome to attitude; it literalizes ethical action. As MacIntyre insists, story is also the context, the background that reveals how the self is inherently embedded in communities, whether he or she is in agreement with, or rebellion against, those communities.[17] As such, the ethical appropriateness of actions cannot be determined without reference to narrative. Put simply, we need to read *The Sound and the Fury* in order to fully understand why the Benjys of the world need to be protected from the Jason Compsons.

This kind of reflection, however, rarely happens. Instead, we cling to autonomy as the highest value, and it compromises our ability to act ethically. I will return to my transhumanist example to illustrate the point. Although Simon Young claims to be writing an ethics of self-enhancement, he defines ethics not as a communal question of what it means to live the good life but only as the question of what to do with our individual lives. His definition of ethics is clearly subservient to his desire for individual autonomy. His answer to the question of how to live the good life is "Live to Evolve! Evolve to Live!" "Evolutionary ethics," he writes, teaches that "we cannot separate ourselves from the ongoing process of evolution—we *are* evolution."[18] So when it comes to the question of how to treat others, technology that encourages evolution is the answer. Since love is "genetically induced benevolence," once we fix our genetic flaws, we will be better at love. "If we are all neuro-chemically empowered, perhaps we may *really* be able to start 'loving thy neighbor as thyself.' As we become able to create the chemistry of enlightenment through biotechnology, perhaps we may *all* come to feel the mystic's bliss of universal love."[19] My point in bringing up this example is not to claim that every American believes technology will eventually provide the neurochemicals we need to learn to love others better. It is not even to argue that neurochemicals cannot work this way. My point, rather, is to ask whether this attitude toward our future helps us to love people better today. And my answer, backed up by the narratives I will present here, is a definitive no. What is more, I will argue that this attitude not only does not help but actively prevents people from becoming persons who are capable of loving one's neighbor as oneself. My thesis boils down to this: when one's ultimate value is freedom to remake the self through technology, this value shapes a view of the other that makes love for *particular persons today* almost impossible. No good decisions about the future can be made from this foundation.

Loving others is admittedly a higher standard than being merely ethical, even if such a thing were possible. Although it has considerable drawbacks, choosing love as the standard has the chief benefit of preventing us from reducing ethics to questions of personal autonomy. Love also has the backing of one of the most powerful ethical formulations of all times: the charge to "love your neighbor as yourself." For centuries Christian theologians have fleshed out this command, arguing through the actions of Jesus Christ that love necessarily involves sacrificing one's freedoms, that it means putting others first when claims are in conflict. But we do not have to call ourselves Christians to recognize the indispensable wisdom of this ancient teaching.[20] At the very least, if we aim at loving others, we will be more likely to act ethically toward them.

Choosing love for one's neighbor as the starting point for ethics has the further benefit of deemphasizing the question of what it means to be a person per se by emphasizing the ethical imperative to love all human others who we must *assume* to be persons. The view that not everyone born into the human community is worthy of the same kind of loving care is sometimes called "personism" and is shared by many of the most influential contemporary ethicists (most notably, Peter Singer) contributing to the debate about how biotechnologies should be used. As this book will illustrate, most Christian philosophers and theologians agree strongly that this position must be rejected. As such, these thinkers share a position that is often termed "personalism": the belief that ethics must start with the basic assumption that human beings are, simply by virtue of being born, persons within the human community and thus our neighbors. Although the Christian thinkers I rely on in this book—Søren Kierkegaard, Robert Spaemann, John Zizoulas, Hans Urs von Balthasar, Paul Ricoeur, and many others—have significant theological differences and come from a variety of traditions, they all draw on Christian doctrines to insist that human beings are defined by a series of relationships and their corresponding responsibilities, first to God and then to others. Although there is a strong tradition of personalism among Catholic theologians, an ethics based on personalism does not require specifically Catholic doctrinal commitments.

One of the other benefits of choosing love as the standard for how persons should treat other persons is that it sharpens into focus the central, if often blurred, question behind most of the thorniest problems in bioethics. That question is: What exactly is the good life for us? St. Thomas Aquinas contended in his *Summa* that "an act of love always tends towards two things; to the good that one wills, and to the person for whom one wills it: since to love a person is to wish that person good."[21] Strangely, when it comes to using technology for human enhancement purposes, we rarely ask the question of what the good life is; we usually just assume we know it. As the philosopher George Grant argues, modern people define goodness in a way that departs from the traditional desire for justice. Goodness had originally been defined in the West as "that which meets us with the overriding claim of justice" and which persuades us that only in desiring justice for others will we "find what we are fitted for." On the contrary, the modern conception of goodness is "of our free creating of richness and greatness of life and all that is advantageous thereto," popularly described as "quality of life."[22] In other words, modern people assume that the good life is defined by things that have little to do with achieving that goodness for others.

When the call is to love one's neighbor as oneself, the question, what is the good life for us?, can reemerge and generate potential answers to the question beyond "that which we can create." For absent deep, ethically motivated inquiry into the nature of the good life, some biotechnological solutions will inevitably have the feel of a doctor saying to a patient, "Has your leg been hurting you? Here, let me chop it off." To lovingly help the one in need, one must know what is truly helpful, and this can be a very complex question indeed. It gets particularly complicated when the one asking for help is asking for her own leg to be amputated. Though it seems bizarre, this issue actually faces many doctors today, as so-called amputee wannabes have been emboldened by the internet to request that doctors cut off arms or legs that the person does not feel to be a part of himself.[23] This issue puts a fine point on the question, can a doctor who begins with the principle of autonomy and a thin vision of the good life know how to respond to a request from a patient to cut off her healthy leg? There are many other issues that biotechnological advances have brought to the fore. Members of the deaf community, for instance, have recently argued for the right to surgically alter their children in utero to ensure that they will be members of that community also. Can anyone without a robust conception of the good life arbitrate these disputes? My point is not to say that these examples have easy answers; they do not. My point is rather that arguing from the value of personal autonomy without a thick investigation into the nature of the good life will not serve us well in answering them.

The question of what it means to live the good life is one that science cannot answer, though many people unwittingly assume it can and does. As we shall see, science applied through technology encourages us to define the good life by the only thing it has to offer: a technological solution to all problems. Albert Borgmann put it best when he explained that in advanced Western societies, "liberal democracy is enacted as technology. It does not leave the question of the good life open but answers it along technological lines."[24] While technology can only unquestioningly follow the cultural definition of what to see as a problem, its availability changes our vision and influences our choices. A bit like the adage that when one is a hammer, everything one sees is a nail, technological society can redefine anything as a technological problem.[25] While it is easy to laugh at the infomercials that depict someone bent over in pain because she doesn't have the newest SuperMop, it is trickier to discern those larger places where the promise of a technological solution is blinding us by its light. I will argue here that technology, by training us to look for quick fixes to whatever ails us, encourages a dangerously thin version of the good life. This vision of the good life has no backbone; it withers at the first sign of adversity and positively collapses under the ultimate pressure of death. For when the good life is assumed to be the life that experiences the least amount of suffering and the maximum amount of happiness, how can we learn to handle the suffering we will inevitably experience? And when the good life is assumed to be the life with the minimum amount of suffering and the maximum amount of freedom from limitations, how can we learn to love others who do suffer and who do experience substantial limitations? If we cannot love people as they are, can we really love? Toni Morrison writes that "love is never any better than the lover." It is time to ask what kind of lovers we are becoming.

THE PROPHETS

The idea of being a prophet strikes most people today the way that being a philosopher struck Mrs. Hopewell in one of Flannery O'Connor's stories, as "something that had ended with the Greeks and Romans."[26] "Prophecy" is not a word that the fiction writers I discuss would necessarily choose to describe their own efforts. But the word (as I am specifically using it here)[27] is apt in both its senses of truth telling and foretelling, as each writer I discuss engages in one or the other or both of these. Put more precisely, though these writers are not all explicitly religious, they all possess aspects of what Walter Brueggemann names the "prophetic imagination." The task of the prophets, argues Brueggemann, "is to nurture, nourish, and evoke a consciousness and perception alternative to the consciousness and perception of the dominant culture around us."[28] Prophets do this first by *criticizing* the culture's "dominant consciousness" and then by *energizing* a people wearied, perhaps even unwittingly, by that dominant consciousness. Artists, of all people, understand the role the moral imagination must play in forging an alternative consciousness in people. If we have been numbed by the dominant consciousness, we need both a jolt to be awakened out of it and the guidance to envision a different way of thinking.

The nine fiction writers I discuss in this book come from several different regions in North America and three different centuries. They each have differing views of literary art, and their religious beliefs range from atheist and agnostic to Catholic and Congregationalist. What unites them is their desire to oppose the dominant consciousness of an advanced technological society, not technology itself. Prophetic criticism, argues Brueggemann, begins with the capacity to grieve. The prophets force us to look at things that we have come to take for granted, and then they argue that we have settled for something far below the truly good life. The prophets of the posthuman unveil and expose three largely unchallenged and invisible cultural assumptions that fuel the biotechnological revolution and make it more difficult to love others: the belief in a protean self with nearly limitless personal freedom, the belief in the glamour and promise of technology, and the dominance of a consumer culture that trains people to buy solutions first. That these three assumptions exist and are problematic is a view shared to a greater or lesser degree by all of the writers I discuss in this book, and the ramifications of each of these assumptions is what the writers flesh out.

THE PROBLEM: THE DOMINANT CONSCIOUSNESS OF AN ADVANCED TECHNOLOGICAL SOCIETY

The Protean Self

Although the transhumanist vision I described at the beginning of this introduction does not directly appeal to most Americans, it is merely a more radical, technologically sophisticated articulation of basic American (and classically liberal) values, including the insistence on personal autonomy, the desire for mobility and freedom, and the belief that the right reasons for doing things must come from the self, not an external authority. Alexis de Tocqueville may have said it best when he wrote that "America is therefore one of the countries in the world where philosophy is least studied, and where the precepts of Descartes are best applied."[29] Most Americans believe that one's soul or spirit defines a person and that success is primarily a matter of attitude and will. We believe that we can shape ourselves, that we can be whatever we want to be and do whatever we want to do. As a result, we get easily frustrated by our limitations and turn to technology to help overcome those limitations—both in ourselves and in others.

To make this case, I will refer to prominent writers whose work illuminates the perceptions Americans have of the self in an advanced technological society. One such writer is N. Katherine Hayles, whose book *How We Became Posthuman* explains how the classically liberal American self has so easily melded with the biotechnological revolution to produce a new category: the posthuman. Hayles argues that Western society became posthuman when the conditions of the Turing test were accepted without debate. Proposed in a paper that Alan Turing wrote in 1950, the test was set up to try to see if people could discern the difference between machine intelligence and human intelligence by "talking" with the other (whether computer or human) through question and answer. If people could be fooled by the computer, Turing argued, it would prove that machines can think. What is important, explains Hayles, is not the answers that the test produced but the way it framed the issue. The test assumed a definition of human life as intelligence, or more specifically, as information patterns—not as embodiment.[30] This definition of life clearly proceeds from a basic Cartesian dualism that separates mind from body, locates everything that "counts" in the mind, and valorizes the subject's freedom and autonomy above all else. The essence of the self is freedom, not limitations or responsibility.

Defining life as an informational pattern has severe implications for sex, race, and other traits specific to our embodiment. The body becomes a mere appendage for our manipulation; as Hayles explains, it is "the original prosthesis we all learn to manipulate, so that extending or replacing the body with other prostheses becomes a continuation of a process that began before we were born."[31] If a person thinks her sex, race, size, or shape is keeping her from her goals, then she can change any of that. The body is plastic; it is there to be reshaped with genetics, operant conditioning, psychopharmaceuticals, or any other enhancement technology available.

Another writer who has convincingly traced the American view of the self with regard to the biotechnological revolution is Carl Elliott in his book *Better Than Well: American Medicine Meets the American Dream.* Elliott explains how the American quest for an "authentic" self is a quest that, ironically, is motivated primarily by how others see us. And the cultural norms are, of course, a shifting target. At the end of the book, Elliott argues that the mere availability of the enhancement technologies is not enough to explain their appeal. "What has made the ground for these technologies so fertile?" he asks. Part of his answer is that "we tend to see ourselves as the managers of life projects that we map out, organize, make choices about, perhaps compare with other possible projects, and ultimately live out to completion."[32] If Americans feel that we are the managers of our own life projects, then we feel responsible for their success or failure, and enhancement technology naturally appeals to us as we try to move forward. Elliott asks, "If my life is a project, what exactly is the purpose of the project? How do I tell a successful project from a failure?"[33] We do not have Aristotle's answers about the telos, the purpose of the human life. We have also largely jettisoned the Judeo-Christian sense of a being made in the image of God.

So can there be, Elliott asks, "such a thing as a single, universal human purpose?" Not, he argues, if we take our cues from the dominant culture. "From philosophy courses and therapy sessions to magazines and movies, we are told that questions of purpose vary from one person to the next; that, in fact, a large part of our life project is to discover our own individual purpose and develop it to its fullest. This leaves us with unanswered questions not just about what kinds of lives are better or worse, but also about the criteria by which such judgments are made."[34] The self is fundamentally its own author in opposition to any potentially limiting relationship with others.

Elliott's argument is hardly new, but it is very important. It is a layperson's version of what many scholars have noted characterizes the postmodern, or late modern, self. The postmodern self has turned away from both medieval notions of providence and enlightenment notions of progress. Brent Waters has made a very convincing case that the change is substantial: in the late modern world, *techne* has replaced *telos;* process has replaced progress. The end result is a world so committed to personal freedom that what an individual wants to change into is far less important than the fact of having the freedom to change.[35]

The Promise of Technology

For the purpose of this book, I am defining technology rather broadly, as science applied through technique to effect some particular purpose. This definition encompasses everything from cosmetics to cosmetic surgery, from the wheelbarrow to the iPod. Clearly, human intervention in nature is nothing new; society has been "technological" since the first person picked up a stick to use as a club. But to say we live in an "advanced technological society" is to say much more than that. Technological growth is rapid and accelerating; advanced Western societies place more confidence in it than ever before. As George Grant put it, "in each lived moment of our waking and sleeping, we are technological civilisation."[36]

Although a fair amount of scholarly disagreement exists regarding the theoretical question of how neutral technology is, contemporary belief in the promise of technology reminds us that the Enlightenment commitment to dominate nature in order to improve the human condition still holds sway.[37] Francis Bacon's *The New Atlantis* contains one of the earliest expressions of the promise in the form of a fictional utopia called Bensalem. In the book, the king of Bensalem tells some admiring English visitors about how far their technology has advanced under the simple foundational principle that "the end of our foundation is the knowledge of causes, and secret motions of things; and the enlarging of the bounds of human empire, to the effecting of all things possible."[38] As Eric Cohen explains, Bacon's vision was part of a general turn toward believing that science applied through technique will be the main way that humankind achieves salvation, which is defined as the amelioration of human misery. "On the isle of progress," writes Cohen, "the priest is replaced by the scientist, who conducts secret experiments to help his fellow citizens. This is the new charity."[39] Albert Borgmann has demonstrated countless ways in which the promise of technology is reiterated and reformulated in American public discourse. One need not believe that technology is itself acting to acknowledge the power of belief in it, for "implied in the technological mode of taking up with the world there is a promise that this approach to reality will, by way of the domination of nature, yield liberation and enrichment."[40] Since the turn of the century, hundreds of books have been published with titles that reiterate this promise. By way of technology we can—just to name a few—challenge nature, enhance evolution, birth babies by design, redesign humans, become more than human, bring about the singularity, transcend all suffering, and conquer death.[41]

Regardless of whether one so explicitly believes in technology as a kind of savior, it is difficult to disagree with Albert Borgmann's central thesis in *Technology and the Character of Contemporary Life,* that technology has shaped our lives by its own particular pattern. This pattern, insists Borgmann, "is neither obvious nor exclusively dominant," but profoundly influences the way we see and interact with the world.[42] Borgmann explains the distinction between thing and device to make the point. A violin is a thing that one must learn to play in order to produce music; a stereo is a device whereby one can only consume music. The replacement of things by devices has proceeded imperceptibly, but it has changed our values in the process. "It is the pervasive transformation of things into devices that is changing our commerce with reality from engagement to the disengagement of consumption and labor. Only if we envision and challenge this inclusive pattern can we . . . discern how we redefine ourselves in the process of implementing the values of technology."[43]

The prophets I discuss in this book do not necessarily agree with Borgmann about how much technology per se has shaped our values, but all of them are concerned with how the promise of a quick fix begins to make us see other persons, particularly other persons whom we claim to want to help.

Purchasing Happiness

Many writers have already demonstrated the strangling saliency of American consumer culture, and I will not take the space to reiterate their arguments. But the biotechnological revolution amplifies the negatives of such a culture because it enables the purchase of enhancements that effect much more significant changes than orthodontics or facelifts. The prophets of the posthuman will reveal how deadly a combination it is to have a protean view of the self coupled with ubiquitous market forces bent on convincing you that their particular technological enhancement is going to give you, finally, the happiness you have always sought. As Carl Elliott explains, "the market moves to fill a demand for happiness as efficiently as it moves to fill a demand for spark plugs or home computers."[44]

While the market strives for constant consumer dissatisfaction, it is also the case that satiety is a particular problem with profound consequences for ethics. Sated consumers in an advanced technological society have a collective consciousness that is similar to Walter Brueggemann's "royal consciousness." Brueggemann compares contemporary America to the ancient empire of Solomon, which was satiated, at ease with itself, and indifferent to the claims of the other. Notably, both Solomonic Israel and contemporary America take well-being and affluence to be the greatest good, regardless of what that might mean for the poor. "Covenanting that takes brothers and sisters seriously had been replaced by consuming, which regards brothers and sisters as products to be used. And in a consuming society an alternative consciousness is surely difficult to sustain."[45] Zygmunt Bauman argues that ethics in a consumer society shifts away from responsibility to others to the realm of self-fulfillment. "The collateral victim of the leap to the consumerist rendition of freedom is the Other as object of ethical responsibility and moral concern."[46] As this book will illustrate, the dominant consciousness of an advanced technological capitalistic society clearly tends toward a utilitarian ethic, an ethic that permits other beings to be used—consumed—when it can be proven to serve the individual *or* the greater good.

If each of these writers is correct, and I believe that they are, what we have in contemporary America is a society of individuals who think that their bodies are essentially plastic, who think of their lives as a project, who look to technology to solve their problems, who value individual autonomy above most other things, and who are encultured to believe that money can buy happiness. This is hardly a robust definition of the good life. We may still value the ideal of "love your neighbor as yourself," but the dominant consciousness is producing people incapable of doing it—or even of thinking about it. So what can be done?

THE PROJECT

Proponents of self-directed evolution through technology tend to characterize novelists who warn about our posthuman future as mere naysayers. These proponents then argue that the prophets are simply fear-mongering in order to sell books, and then they use that negativity as a cause to dismiss them. But as I mentioned earlier, Brueggemann explains that to be a prophet is to criticize and to energize, and this book will reveal how each of these writers is engaged in both activities.

Because the primary issues in ethics involve the question of how we see others, part 1 of this book, "Posthuman Vision," describes where the dominant ethical vision in technoscientific writing comes from and what it leads to. In the first chapter, "The Moral Imagination in Exile:

Flannery O'Connor and Lee Silver at the Circus," I will demonstrate how current scientific writing (the kind that forays into ethics)[47] has shortchanged the role of the moral imagination. This move has not only negatively affected how individuals learn to see others, especially people with disabilities, but also undervalued the role of the literary in decision making. I will then offer Flannery O'Connor's short story "A Temple of the Holy Ghost" as an example of her attempt to expand our vision of the good life to include a greater responsibility for others who do not easily fit into categories supplied by scientific naturalism. In chapter 2 I will show how Nathaniel Hawthorne's short story "The Birth-mark" adds to the critique by illustrating how a reliance on technology keeps people in moral infancy, promoting a narrow vision of what love for the other can be.

Since human enhancement also entails a certain commodification of the human body that has ethical ramifications, part 2, "Posthuman Bodies," deals with various narratives that challenge that view of the body. In chapter 3 I turn to the beauty industry's easy dalliance with technology and consumer culture. Through shows like Fox's *The Swan,* in which "ugly duckling" women are given a full array of cosmetic surgeries and then brought into a beauty pageant to compete against each other, technology is promoted as the best chance for people's happiness. The stories of James Tiptree Jr. and George Saunders show how consumer culture contributes to a numbing failure to love others, a failure that becomes more pronounced the more technology is relied on to produce it. In chapter 4 I will argue that although Toni Morrison's novel *The Bluest Eye* is not about technology at all, through it we can ponder the eventuality that one day girls like Pecola Breedlove—a little black girl who wants blue eyes—will have access to technology that would give her those blue eyes and virtually any other physical feature she desires. The question that Morrison forces us to address is just what such a promise of "improvement" does to a little girl like Pecola. Morrison's novel is adept at revealing how people who claim to love others the most may be hurting them most deeply of all. Morrison's novel will also serve as a good transition to the dual purpose of these writers to criticize and energize, as Morrison clearly offers an alternative consciousness to the dominant consciousness in the character of Claudia MacTeer. Since

Claudia is eventually able to step outside of her community's paradigm for seeing others, she becomes the locus of hope for Morrison.

The Bluest Eye also reveals how much Morrison has in common with Margaret Atwood and Walker Percy, each of whom believes that our impoverished language is part of the problem in the dominant consciousness. Part 3, "Posthuman Language," demonstrates how, since our ethical decisions are compromised by thin thinking about the nature of the good life, a sufficiently thick language must be recovered if we are to learn what the good life truly entails. In chapter 5 I will reveal how Atwood's speculative novel *Oryx and Crake* demonstrates that debased language deforms the moral imagination, helping to create people who are capable of putting their own visions of perfection above even the very survival of humanity. Walker Percy similarly blames the abstraction "love for Humanity" for many of our worst attitudes toward helping others. His novel *The Thanatos Syndrome,* the focus of chapter 6, both criticizes the dominant consciousness and energizes an alternative view by providing rich and embodied language of healing, hospitality, and humility.

With this focus on healing, hospitality, and humility, Percy's novel also serves as an example of the other part of what the prophets offer: a fully imagined alternative consciousness. Thus Percy's novel is a good transition to part 4, "From Posthuman Individuals to Human Persons," in which I describe two writers who point the way back to the finite and dependent human person from our exile into the illusion of self-determination. Raymond Carver's story "A Small, Good Thing" reveals how the technological society has dangerously limited the possibility for experiencing and learning from contingency, which is needed for people to receive grace and love others. His story shows how new vistas can be opened by simple acts of hospitality. And Marilynne Robinson's *Gilead,* the most recent of these novels, brings readers back to a world where one's approaching death is the catalyst for personal moral growth that can be gained in no other way. Robinson's protagonist, the pastor John Ames, is able to learn to love even the most challenging people in his life in large part because he understands his own finitude, limitations, and dependence on others. His story reveals how important what Alasdair

MacIntyre calls the "virtues of acknowledged dependence" are to human identity and moral agency.[48]

Each of the nine creative writers I engage with in this book believes in the power of art to help us to envision the good life fully. Art is needed not only to show the way to a richer reality but to make that reality more possible by offering it through a parallel experience. As the poet Denise Levertov put it, the revelations out of which prophecy and poetry come stir the prophet or the poet to expressions designed for others to experience: "they breathe in revelation and breathe out new words; and by so doing they transfer over to the listener or reader a parallel experience, a parallel intensity, which impels that person into new attitudes and new actions."[49] These writers believe in the power of art to redirect readers' thoughts and actions. Without that belief, they would not have chosen to be writers at all.

Though it has been drawn on perhaps too often to see afresh, the biblical parable of the good Samaritan provides irreplaceable insight into the power of prophetic narrative to redirect thinking about the ethics of personhood.[50] The narrator of the Gospel takes care to describe the scenario within which Jesus told the parable, explaining that a lawyer was trying to put Jesus to the test by asking him what he needed to do to inherit eternal life. Jesus responded, as he often did, with a question: "What does the law say?" The lawyer answered him, "You shall love the Lord your God with all your heart and with all your soul and with all your strength and with all your mind, and your neighbor as yourself." Jesus told him that he answered correctly; "do this, and you will live." But instead of leaving it at that, the lawyer tried to justify himself by asking, "and who is my neighbor?" The most telling thing about this scene is the way in which Jesus does not answer this question. He does not define "neighbor" by delineating which other people the lawyer should set out to love. Instead, he tells a story about a Samaritan, the only one of three passersby who had compassion and bothered to stop to help a man who had been attacked by robbers. And then he asks the lawyer, "Which of these three, do you think, proved to be a neighbor to the man who fell among the robbers?" Jesus turns the tables on the lawyer, zeroing in on his beliefs and behavior. Although the lawyer wants to escape responsibility by deciding who or what a person is or is not, who should be called "real," or who exactly is his neighbor and why, the parable prevents him. What the lawyer must learn is that he *already is a neighbor*. What he must learn first is not *who* to love, but *how* to love. Love then dictates the appropriate recognition of the other.

The move that Jesus makes in this parable thus can also serve to describe the ethics of personhood that lies behind this book. Christian thinkers who have contributed to, or drawn from, personalism (as I have used the term above) agree that persons are defined not by their specific manifestation of certain qualities, but by relationship with others that entails the idea of the person as a gift. The Samaritan gave himself to a stranger for whom he had compassion for no other discernible reason than that he saw that the person was in need. The Samaritan proves that he is capable of recognizing and loving other persons, even when it involves considerable risk and no clear gain. The fact that the Samaritans were considered outcasts by Jesus's original audience for this parable makes the point even more sharply. The call to love your neighbor as yourself is a severe challenge to personal bias. This is why Jacques Maritain so often insisted that while individuality is a simple material issue, personhood is a spiritual one that involves mystery. Individuals can be described by their bodies and their desires, but persons cannot be described without reference to love. What we love when we love other persons is "the deepest, most substantial and hidden, the most *existing* reality of the beloved being," and this is a "metaphysical center deeper than all the qualities and essences which we can find and enumerate in the beloved."[51] When we love this way, we also ourselves enter more deeply into the mystery of personhood.

The fact that persons exceed their "qualities and essences" is the main reason why ethics requires narrative. Through a narrative Jesus showed the lawyer that his concern should be how he could become the good Samaritan himself. His concern should be how he can become the kind of person who is defined by love, who sees beyond his own existence into the reality of another person. He must learn to answer the question that lies underneath the parable but is not answered by it: How was it possible that the Samaritan was able to have compassion on his neighbor and act accordingly? And so the lawyer is himself on a journey, assigned to him by Jesus: "Go and do likewise." The parable has begun that journey for the lawyer, but it is up to him to continue it.

The prophets of the posthuman understand that the journey to becoming this kind of loving neighbor cannot be transcended or hurried. Simply living life—no matter how long, pain-free, or enhanced that life is—does not necessarily put a person on the path toward loving his neighbor. If we suppose that our neighbor is there and that she is real, it will be an arduous journey toward learning to love her, a journey that each person and every generation must learn for themselves. If there is one thing each of these writers agrees on, it is that there are no technological shortcuts.

PART
I

Posthuman

Vision

CHAPTER 1

The Moral Imagination in Exile

Flannery O'Connor and Lee Silver at the Circus

> *Writing can be a formal way of enacting Oliver Cromwell's plea:*
> *"I beseech you, in the bowels of Christ, think it possible you*
> *may be mistaken."*
>
> —George Saunders

Of all of the contemporary thinkers trying to describe the quagmire of the current ethical landscape, none may be more revealing than those who call for a new science of morality. Sometimes calling themselves "evolutionary psychologists," sometimes "sociobiologists," and sometimes just "Darwinian naturalists," this group of thinkers, ranging from Daniel Dennett to Steven Pinker to Robert Wright, shares a desire for a "consilience" between biological research and cultural studies. They believe that once everyone submits to the "undisputable fact that we human beings are products of evolution," new ethical norms will emerge that will provide us, finally, with empirically based answers to the question, how should we then live?[1]

Though there is some disagreement among these thinkers, they share enough to be invited to the table in one of the hottest clubs in town: the Edge Foundation. This group, founded by John Brockman, is so called because that is where they want to be: on the cutting edge of a new consilience between science and culture. Following the lead of E. O. Wilson, Brockman champions the creation of a "third culture," which he describes as consisting "of those scientists and other thinkers in the empirical world who, through their work and expository writing, are taking the place of the traditional intellectual in rendering visible the deeper meanings of our lives, redefining who and what we are."[2] Brockman explains that, in 1975, E. O. Wilson predicted that "ethics would someday be taken out of the hands of philosophers and incorporated into the 'new synthesis' of evolutionary and biological thinking. He was right."[3]

In July 2010, the Edge Foundation organized a conference to discuss and develop this new science of morality, and its proceedings are illustrative of the range of approaches permitted by those invited to take the place of the "traditional intellectual." The major presenters came from the fields of neuroscience, biology, psychology, and experimental philosophy. These presenters share a naturalistic approach in which empirical data, taken from either brain research, psychological experiments, or sociological studies, is the only acceptable starting point for ethical reasoning. These scientists claim to be aggressively changing the field of moral psychology by using "babies, psychopaths, chimpanzees, fMRI scanners, web surveys, agent-based modeling, and ultimatum games" to remap the behavioral sciences. Among the presenters Brockman introduces are Joshua Green, a cognitive neuroscientist and philosopher at Harvard who argues that "our brains trick us into thinking that we have Moral Truth on our side when in fact we don't, and blind us to important truths that our brains were not designed to appreciate," and Yale psychologist Paul Bloom, who argues that "humans are born with a hard-wired morality. A deep sense of good and evil is bred in the bone."[4]

There is nothing inappropriate about any of this research. These scientists are breaking important new ground. What is problematic for ethics is the attitude that many of these third-culture scientists tend to take toward disciplines such as philosophy, theology, and literary studies. In his introduction of the group, Brockman clearly postulates an "us

versus them" struggle between those who start with biological founda-
tions and those who do not. According to him, the new third-culture
thinkers are winning the battle: "Scientists engaged in the scientific
study of human nature are gaining sway over the scientists and others
in disciplines that rely on studying social actions and human cultures
independent from their biological foundation."[5] By going with Flannery O'Connor and Lee Silver to the circus, this chapter questions the
Edge Foundation's assumption that empirical facts must be the starting
point for moral and ethical reflection. I will demonstrate how in practice that assumption produces not a new consilience of science and
culture, but an oppressive rhetorical framework in which the moral
imagination is exiled. The result is that the person as person disappears
from view, making ethics difficult, if not impossible.

DARWIN'S DANGEROUS IDEA

Most evolutionary psychologists make no effort to hide their naturalist assumptions or their desire to change the ethical landscape. Daniel
Dennett, in *Darwin's Dangerous Idea,* states it clearly:

> Let me lay my cards on the table. If I were to give an award for the
> single best idea anyone has ever had, I'd give it to Darwin, ahead of
> Newton and Einstein and everyone else. In a single stroke, the idea
> of evolution by natural selection unifies the realm of life, meaning,
> and purpose with the realm of space and time, cause and effect,
> mechanism and physical law. But it is not just a wonderful scientific
> idea. It is a dangerous idea.[6]

The evolutionary story appeals to Dennett because of its explanatory
power: it unifies everything under a physical explanation. In effect, we
no longer have to worry about the deep truths of existence in the gut-
wrenching way we used to. Instead, we can spend our time turning to
the natural world, studying it empirically, letting it be our guide to explaining "life, meaning, and purpose." Dennett explains how he thinks
ethical principles should be derived: "Ethics must be *somehow* based on

an appreciation of human nature—on a sense of what a human being is or might be, and on what a human being might want to have or want to be."[7]

In theory, Dennett's belief that "ethics must be *somehow* based on an appreciation of human nature" is a position open to a productive consilience between science and the humanities. After all, "human nature" is a big topic, and to find out what a "human being might want to have or want to be" should warrant an invitation to traditional philosophers, literary scholars, and even theologians. But in practice, no such invitation is extended, and "Darwin's dangerous idea" becomes dangerously exclusive. At the end of the day, the new science of morality permits only empirically verifiable accounts, rhetorically bullying anyone who disagrees with the choice. Here is one example of how this rhetoric operates. Each January since 1998, the Edge Foundation's "world question center" has come up with a provocatively general question that it asks various people to answer. In 2008, Edge sent this question to one hundred and fifty handpicked respondents: "What have you changed your mind about? Why?" The respondents were given this prompt written by John Brockman: "Science is based on evidence. What happens when the data change? How have scientific findings or arguments changed your mind?"[8] Before anyone can offer answers, they are hit with the following preamble, representative of the basic assumptions of the group:

> When thinking changes your mind, that's philosophy.
> When God changes your mind, that's faith.
> When facts change your mind, that's science.[9]

While this division may seem innocuous, it is a typical third-culture rhetorical move implying a clean distinction between facts and anything having to do with faith, as if anything that comes from faith is automatically counterfactual, unprovable, merely spiritual, or all three.[10] It also implies that everything we need in order to inhabit our world together is demonstrable by science. The rest is immaterial.

This shows that there is not much new about the new science of morality. Marilynne Robinson's book *Absence of Mind* illustrates how writers like Daniel Dennett and Steven Pinker share naturalistic assumptions that actively shape the conclusions (particularly about the nature of human consciousness) that they claim to be merely scientific. Their work begins, she argues, with an unproven modernist conclusion that "positivism is correct in excluding from the model of reality whatever science is (or was) not competent to verify or falsify." Robinson then argues that this excludes "virtually all observation and speculation on this subject that have been offered through the ages by those outside the closed circle that is called modern thought." Specifically, it excludes any true subjectivity, the idea of one's "felt experience." The new thinkers, whom Robinson calls "parascientists," want instead to "solve" questions "in order to further the primary object of closing questions about human nature and the human circumstance."[11]

Robinson provides an interesting example of this tendency to close questions in her account of the way the story of Phineas Gage is typically handled by evolutionary psychologists. Gage is the railroad worker who, in 1848, did not even lose consciousness when an iron rod pierced his skull and damaged parts of his brain. Gage survived but experienced a complete change in personality, becoming surly and aggressive. Gage's story and ones like it are assumed to prove naturalist assumptions about the brain as the ultimate seat of morality. In other words, the only facts that matter about Gage's case are the facts of his brain injury and how it affected his behavior. Robinson does not deny that Gage's brain injury is what caused his personality to change, but she questions how reductively the anecdote is used:

> I trouble the dust of poor Phineas Gage only to make the point that in these recountings of his afflictions there is no sense at all that he was a human being who thought and felt, a man with a singular and terrible fate. In the absence of an acknowledgement of his subjectivity, his reaction to this disaster is treated as indicating damage to the cerebral machinery, not to his prospects, or his faith, or his self-love. It is as if in telling the tale the writers participate in the absence of compassionate imagination, of benevolence, that they posit for their kind.[12]

Marilynne Robinson is drawing attention to examples of a failure to acknowledge the role in ethical decision making of the moral imagination. "Moral imagination," as I am using the term here, is the faculty engaged in interpreting and valuing the facts of the world as they appear to the moral actor. In the example of Phineas Gage, the moral imagination is discounted both in Gage himself and in the scientific naturalist's interpretation of the facts surrounding him. In other words, the assumptions of scientific naturalism produce a kind of blindness to the larger human story, a blindness that prevents both an adequate account of morality and an adequate view of the moral actor himself.

With this observation, we are not far from the work of the mid-twentieth century philosopher-novelist Iris Murdoch. Murdoch attacked scientific naturalism as a starting point for moral philosophy.[13] In "Vision and Choice in Morality" Murdoch argues that naturalistic assumptions ignore the fact that "we differ not only because we select different objects out of the same world but because we see different worlds."[14] In other words, moral agents—like Gage and the scientists who interpret his story—do not make choices based on simple facts; they make them according to a particular vision of the world, a perspective, a Gestalt. Murdoch explains that as soon as "we attend to the more complex regions which lie outside 'actions' and 'choices,'" then we can recognize that moral differences stem from something much deeper, from differences of understanding, "which may show openly or privately as differences of story or metaphor or as differences of moral vocabulary betokening different ranges and ramifications of moral concept."[15] By arguing that the basic assumptions of scientific naturalism prevent the vocabulary necessary for understanding morality, Murdoch does not challenge science, but she does challenge, as one critic puts it, the "worldview that places the natural sciences in the position of dictating the whole of our metaphysics."[16] Murdoch and Robinson would agree that Phineas Gage is a complex moral actor; mapping changes in his brain will never fully account for his behavior.

One fascinating irony is that some prominent neuroscientists are now corroborating Murdoch's description of moral vision, story, and metaphor. Mark Johnson and George Lakoff helped to begin a revolution in cognitive neuroscience by advancing a theory that insists that the mind cannot be disembodied.[17] In his book *The Moral Imagination,* Johnson continued his research in neuroscience, but this time with a focus that has immense implications for the basic assumptions of scientific naturalists. He argues empirically for the centrality of the moral imagination, redefined with a neuroscientific twist:

> We human beings are imaginative creatures, from our most mundane, automatic acts of perception all the way up to our most abstract conceptualization and reasoning. Consequently, our moral understanding depends in large measure on various structures of imagination, such as images, image schemas, metaphors, narratives, and so forth. Moral reasoning is thus basically an imaginative activity, because it uses imaginatively structured concepts and requires imagination to discern what is morally relevant in situations, to understand empathetically how others experience things, and to envision the full range of possibilities open to us in a particular case.[18]

Simply put, and contrary to conventional wisdom, moral reasoning is less dependent on laws or principles than it is on metaphorical concepts. As such, "the way we frame and categorize a given situation will determine how we reason about it, and how we frame it will depend on which metaphorical concepts we are using."[19] This is strong language: the metaphors we choose to live by are what give us the frame, and the frame determines how we reason about any given situation. The larger question of where those metaphors come from, or their validity, is one that cannot be exhaustively answered by empirical methods.

This conclusion is essentially no different from what Murdoch insisted on and Robinson reiterates. Murdoch and Robinson are novelists, and novelists know that Mark Johnson is basically correct: no one makes moral judgments on the basis of empirically verifiable facts alone. While brain chemistry may predispose someone to seeing a certain thing a certain way, it neither dictates nor structures that choice. This must alter the assumptions of scientific naturalists. To say that "when the facts change your mind, that's science" is meaningless because the same facts can change two people's minds in completely different ways. The day-to-day translation of experience into meaning, and of meaning into moral decisions, is complex and profound. That translation is conducted through the imagination, and imagination is affected by values we do not even know we possess and by beliefs we thought we had abandoned.

The point of this chapter is not to argue against the important work that neuroscientists and others are doing to understand the physiology of morality. It is instead to challenge the assumption that those physiological facts are relatively easy to interpret and should supersede any other type of reasoning. Novels show that neither moral reasoning nor reasoning about morality escapes embodied, imaginative, and lived interpretation of facts and experiences. These interpretive schema cannot be empirically verified; they can only be tried on. Embodied narrative is the only way to arbitrate between available interpretive possibilities, and when two different moral perspectives are actually tried on, the result is clarifying. The naturalist's lens tends to filter out whatever and whomever it cannot explain, but the writer whose moral imagination has not been delimited by these assumptions often sees more. To illustrate, I will put two stories side by side; both tell of encounters with a circus "freak." Biologist and Edge-type intellectual Lee Silver's account comes from his *Challenging Nature,* and Catholic novelist Flannery O'Connor's from her story "A Temple of the Holy Ghost." While Silver is certain the empirical facts of an encounter with a circus freak speak one way, O'Connor's story directly challenges that perspective, forcing the onus back on the viewer. This difference illustrates the importance of the moral imagination—and by extension, the world of literature—to thinking about ethics.

TWO WRITERS AT THE CIRCUS

Lee Silver is an effective prose stylist, and his book *Challenging Nature: The Clash of Science and Spirituality at the New Frontiers of Life,* is calculated and fascinating. A mix of personal narrative, sociological argument, and scientific description, it is a compelling apologetic for biotechnology. He begins chapter 9, "Counting Souls," with two contradictory epigraphs. The first is from Francesco A. Lentini, who was born in 1899 with three working legs and two sets of genitalia. Lentini said that "my mother gave birth not to two children, but more than one, yet not two." The second quotation is from Charles T. Rubin, a professor of political science at Duquesne University: "We [human beings] don't have three hands, and being two-handed creatures may be significant for living fully and truly as human beings. We are embodied in a particular way."[20]

The epigraphs set up what Silver has already argued is the main issue in the debate over the use of biotechnology to enhance human beings: namely, the question of whether or not a firm line can be drawn between human and nonhuman. Here, Silver believes that he defeats Rubin by exposing his commitment to a particular view of human nature that excludes people like Lentini. Silver then provides multiple examples of people born with genetic deformities, including Josephine Myrtle Corbin, whose spine bifurcated into two and led to the growth of two pairs of legs and two sets of sexual organs (out of which she had a total of five children), and a man known as Two-Faced Chang who had an additional face growing out of his right cheek. Silver also describes examples of the rare occurrence of *fetus-in-fetu,* when a living fetus grows inside of another fetus, and examples of split-brain human bodies, in which one body clearly has two heads, each with his or her own separate mind. Each step along the way he uses these facts to argue that no stable definition of human can exist and that therefore the mind "emerges from the physical brain, not from a spiritual dimension." Silver believes that since a hard and fast line between human and nonhuman cannot be drawn, decision makers are forced to leave "emotionality" and "its cousin, spirituality" out of the equation.[21]

The analysis that follows will show that Silver believes the following: first, the only rational interpretation of the physical evidence is that there is no separable soul; second, since that is the case, the body can be altered without moral problem by anyone who chooses to do so (and, for example, the growing of headless humans for organ replacement is morally acceptable); third, anyone who argues to the contrary is making an emotional or spiritual argument and not a scientific one; and fourth, since people have different emotions and spiritual beliefs, those cannot be relied on to arbitrate disagreements. In spite of what he claims, the interpretation of the facts Silver presents is, in fact, utterly dependent on a value system and imaginative schema. This can be seen as Silver narrates his own youthful encounter with a traveling circus and its presentation of a baby born with two heads.

It was 1964, and Silver remembers that the circus barker who was trying to sell the sideshow would yell out, "Baby with two heads; born alive, stayed alive." Unwilling to part with his dime too easily, the precocious Silver asked the barker if the baby was *still* alive. The barker ignored him, but Silver decided to risk his money "to see this baby whose image in my mind had shaken my notions of what it meant to be a human being."[22] Silver narrates the event with great skill, describing the darkness of the room, the smell of the formaldehyde, and the illumination of the jar containing the baby. Others left quickly, but he stared for a long time.

> Was it real? My youthful scientific instincts told me that the humanlike features seemed too perfect to be otherwise. The fingernails, toenails, and properly positioned wrinkles on each digit were finely drawn; and the eyebrows, all four, were sculpted with perfectly aligned hairs. Perhaps the barker had not lied. The baby may have been born alive. But how could one body contain two persons, I wondered, and if there was only one person, in which of these two perfect heads had it existed? The unspoken question hastened my deepening disillusionment with faith-based absolutist notions of human individuality.[23]

The fact of the baby caused the young Silver to question his previous assumptions. Silver could now plainly see that it is possible for a baby to be born with two heads. While Silver would suggest his reaction was without emotion, the scene itself belies him. Although he could have left, he stayed, filled with wonder. This sense of wonder leads him to a metaphysical question: Was there one person here, or two? This is not a question that the facts themselves can answer. But instead of leaving the experience open to different interpretations, Silver subsumes it into his own narrative of "deepening disillusionment" with certain people of faith. He lumps together "faith-based" and "absolutist no-tions," as if to have faith is the same as fundamentalism.

Silver is wrong to argue that the facts speak unequivocally here. He cannot prove that his response was not emotional or even spiritual. It is decidedly a metaphysical response to believe that this event proves that there are no souls. The fact that it is possible to imagine a different, but equally rational, response to the same facts reminds us that emotions, spiritual convictions, and moral values shape *all* responses. Further-more, if Mark Johnson is correct that moral imagination is shaped by metaphor, the metaphor informing Silver's conclusions here is that of the scientist encountering a laboratory specimen. Silver applied his "youthful scientific instincts" at the time, instincts that his adult imagi-nation fully incorporated into his moral frame of reference. The meta-phor of the laboratory leads to a narrative frame that permits only certain types of questions to be asked about the circus freak, namely, "is it one person or two, and how can we tell?"

In Flannery O'Connor's story "Temple of the Holy Ghost," the protagonist is an unnamed child on the brink of adulthood, intelligent, and just as full of questions as the young Silver. She, too, is confronted by a circus sideshow displaying something that she cannot fully assimi-late, something that challenges her definition of what it means to be human. She encounters a hermaphrodite.[24] But, in O'Connor's hands, the display of an "accident" in nature increases the wonder and awe of the perceiver, rather than her disillusionment. And so O'Connor's tale defends the role a moral imagination shaped by faith plays in decision making, and offers a way that imagination can guide encounters with other human beings.

O'Connor's choice in narrative perspective is illustrative. The story has a third-person omniscient narrator who follows a twelve-year-old girl referred to only as "the child." The child is clearly in the midst of making all kinds of important decisions about who she is, how she sees the world, and particularly, how she sees others. Like many precocious children, she seems to default to a judgmental rejection, mocking every-thing and everyone she deems inferior. From the start, she has rejected her second cousins Joanne and Susan; she "decided, after observing them for a few hours, that they were practically morons and she was glad to think that they were only second cousins and she couldn't have inherited any of their stupidity" (*CW,* 197). The child sasses her mother, makes fun of Miss Kirby and her admirer, Mr. Cheatham, and even prays self-righteous prayers after meeting some "stupid" local boys: "'Lord, Lord, thank You that I'm not in the Church of God, thank You Lord, thank You!'" The only one who ever calls her on her bad behavior is the cook, who asks, "Howcome you be so ugly sometime?" and reminds her that "God could strike you deaf dumb and blind . . . and then you wouldn't be as smart as you is." The child's response is typical: "I would still be smarter than some" (*CW,* 203).

Like so many of O'Connor's protagonists, the child runs headlong into something that she cannot so easily assimilate, in this case, the circus hermaphrodite. Because she is not old enough to go with her cousins to the circus, she tricks them into giving a full report of the "you-know-what" that they had seen there:

> It had been a freak with a particular name but they couldn't remember the name. The tent where it was had been divided into two parts by a black curtain, one side for men and one for women. The freak went from one side to the other, talking first to the men and then to the women, but everyone could hear. The stage ran all the way across the front. The girls heard the freak say to the men, "I'm going to show you this and if you laugh, God may strike you the same way. . . . I never done it to myself nor had a thing to do with it but I'm making the best of it. I don't dispute hit." (*CW,* 206)

O'Connor's economy of style dictates that readers pay attention to everything. Here it is noticeable that the cousins care more about their experience of the hermaphrodite than they do about him as a person, as they promptly forget his name. But the child's imagination is activated by the story, and as she lies in bed, she tries desperately to picture a person who could be a man and a woman both. In her mind, the scene transforms into a church scene, with the people responding, "Amen, amen," every time the hermaphrodite says, "God done this to me and I praise him." In the child's inner vision, the hermaphrodite's sermon incorporates the things she had been thinking about earlier: "Raise your-self up. A temple of the Holy Ghost. You! You are God's temple, don't you know? Don't you know? God's Spirit has a dwelling in you, don't you know?" (*CW,* 207).

Although the factual event is similar to the young Lee Silver's experience, the child's response is completely different. In the face of something she can neither understand nor explain, she tries out certain bits of exactly the kinds of "faith-based notions" that Silver had summarily rejected. Specifically, she begins to develop a deep empathy for the hermaphrodite; she begins to understand that she is not so different from him. Earlier, the cousins had been mocking their St. Scholastica education that instructed them to resist sexual advances by lifting a hand and saying "Stop sir! I am a Temple of the Holy Ghost!" The child did not think this funny; she was pleased with the phrase, as it "made her feel as if somebody had given her a present" (*CW,* 199). As the idea of all people being the resident places of God begins to swell in the child, her moral imagination is activated and further shaped. The idea of the hermaphrodite expands her view of life as a gift. Her sense of wonder, compassion, and humility all increase.

Since the child's development is O'Connor's primary interest, the story does not end with this vision. When the child and her mother take the girls back to the convent, the reader sees how much the child's vision has begun to change. At Mt. St. Scholastica she finds herself overwhelmed by the congruence of worship and just the sort of acceptance of the wonder of human difference she had been mulling over. A "big moon-faced nun" swoops in on her, the kind of nun who "had the tendency to kiss even homely children," rejecting no one. The service begins, and when the priest kneels to raise the monstrance (the vessel in which the Eucharist is "demonstrated" during ceremonies like this one, the benediction of the holy sacrament), the child thinks of the tent at the fair saying, "I don't dispute hit. This is the way He wanted me to be" (*CW,* 208–9).

This scene is complex. O'Connor brings together a number of aspects of her Catholic faith: the mysteries of the Eucharist and the Incarnation, the idea of humanity being made in God's image, and the idea that creation is a divine gift. O'Connor uses the word "mystery" theologically. In the ordinary sense of the word, "mystery" connotes that which is unknown; in theology, it can mean a truth that has been revealed by God.[25] The difference is vital, for when Edge-type thinkers meet the idea of mystery, they may dismiss it with a God-of-the-gaps argument: that the divine is simply a substitute explanation for what science has yet to explain. O'Connor means something else entirely. For her, the mystery of the Eucharist, the mystery of Christ, and the mystery of the hermaphrodite are connected because God has been *revealed* in all of these things. Mystery is the unveiling of meaning, not the guarding of an inexplicable secret. The fact that this child sees the great mystery of human life is the point of the story. When the child enters worship, sees the monstrance being lifted, and thinks about the hermaphrodite, she senses that she is in the presence of God. As such, she knows that a response is in order, and she begins to pray. Her prayers are very telling: "Hep me not to be so mean. . . . Hep me not to give her so much sass. Hep me not to talk like I do" (*CW,* 208). She takes a look into herself and recognizes that she had been judgmental of others instead of filled with the wonder of being. Taught by the church, the child in O'Connor's story moves one step closer to the acceptance of the bizarre variety of human otherness that is required if one is to learn to love one's neighbor as oneself.

To lay the narrative of "Temple of the Holy Ghost" beside Silver's story is to insist that the view we have of others is both a product of the moral imagination and at the heart of ethics. Both the child and the young Silver wrested the "freak" from its designation as a spectacle in the fair, but both of them still had to see the being through some new schema. *It is not possible to see just the facts.* The key difference here is metaphor. The metaphor guiding Silver's imagination was that of a specimen in a lab; the one guiding the child's was that of the body of Christ in the Eucharist and in the world of persons.

What the two narratives also suggest is that to move toward the perspective of the scientific naturalist correlates with a tendency to see humanity in a smaller range, not a wider one. The perspective makes the circus freak less and less human. This narrowing range of what qualifies as human can be seen in the reasoning of Peter Singer, the Princeton bioethicist. Singer understands that to be able to support embryonic research and abortion, one must have a view of the person that is tied to whatever qualities and capacities he or she currently possesses, not capacities that he or she will possess, or capacities that the species to which he or she belongs possesses. Reasoning further from this position, he supports the right of a parent to kill a mentally incapacitated child under a certain age. He then argues that even in the origins of Western civilization, "infants had no automatic right to life. Greeks and Romans killed deformed or weak infants by exposing them to the elements on a hilltop." Present attitudes about the value of human life, writes Singer, arrived only with the coming of Christianity. Singer is consistent. He argues that if reason alone is our arbiter, there needs to be some specific reason why a potential person should have the same rights as a person does, "but what could that reason be?" He goes on to dismantle many possible reasons, concluding, "Is there any other significance in the fact that the fetus is a potential person? If there is, I have no idea what it could be."[26]

Singer has exiled as invalid the kind of feeling—in this case, empathy or even basic respect—that makes most people judge any child as worthy of protection and care. The moral imagination, with its insistence on a variety of choices for reigning metaphors and interpretive schema, has been excluded, but this exile shapes the vision of the seer just as much as any religious view does. Thus, to return to the circus, Lee Silver's banishment of any ideas he deems faith-based is not fundamentally different from the child's acceptance of the idea that the human body is a temple of the Holy Ghost. Both Silver and O'Connor (through the child) see an example of an unusual human life; one sees a specimen that disproves the religious idea of the soul and moves the seer toward skepticism; the other sees the wonder of God in human form and moves the seer toward love. The question of whose view is correct simply cannot be answered by reason alone. As Eric Cohen analogizes, "reason alone did not teach us slaves were men."[27]

A final example displays the role of the moral imagination in bioethics. Earlier in Silver's *Challenging Nature,* he tells the story of an undergraduate at Princeton University who approached him with an idea for her senior thesis. In class that week, Silver had been lecturing on the similarity between chimp and human DNA, explaining how many scientists believe that with a few genetic adjustments, a human/chimp hybrid could probably be born. The young woman went into Silver's office and told him that she wanted to do it. "I want to combine one of my eggs with chimpanzee sperm and follow the development of the embryo inside my uterus. Then I'd write up my observations for my senior thesis." Dumbfounded, Silver asked what she would do with the baby after it was born, saying, "If it were a chimp, you might donate it to a primate research center or a zoo. But it wouldn't be a chimp, would it? If it were a human child, you would have to treat it like a baby and decide whether to raise it yourself or put it up for adoption. . . . But it wouldn't be exactly human either, would it?" The woman was flustered; she had obviously not thought this much in advance. "I guess I was thinking that I would abort it right before it was born, because my senior thesis would be done, and I'd want to finish the experiment and graduate. . . . So, Professor Silver, what do you think?"[28]

Silver explains that he was "appalled at the time" but couldn't figure out exactly why. He certainly could not draw on his naturalistic view of embryos to argue against this idea. When he asked his colleagues, they had similar negative reactions, and none could offer a full slate of reasons why she should not go forward. My point in raising this example is to highlight the scenario's ethical weight. Silver was clearly in a position of influence in this young woman's life; what he thought and what he said mattered. His ability to model responsible ethical thinking mattered. His moral imagination mattered. I read eagerly to the end of the chapter, hoping to hear what Silver told the student. He doesn't say. Evidently she abandoned the project.

Whether we want them or not, these kinds of bioethical decisions are upon us. Arbitrating between all possibilities will be disastrous if we deny that individuals make decisions through a complex tangle of facts, emotions, spiritual beliefs, and values. Since the biotechnological revolution involves the most fundamental ethical questions of the purpose and value of human life, it is time to attend to who answers those questions and how.

CHAPTER 2

Aylmer's Moral Infancy

Nathaniel Hawthorne and the Quest for Human Perfection

> *Liberal democracy is enacted as technology. It does not leave the question of the good life open but answers it along technological lines.*
>
> —Albert Borgmann, *Technology and the Character of Contemporary Life*

It is easy to think of enhancement technology in terms of bold changes to the body that are not yet fully possible: implants to help soldiers see at night; gene therapy to make athletes' bodies more efficient; drugs that enable college students to go for days without sleep.[1] Though these bioenhancements move users "beyond species norms,"[1] less radical enhancement technologies have existed for years and may be much more indicative of general attitudes toward biotechnology than we recognize.

Consider the following scenario. It is May, and two neighbors, good friends, happen to see each other as they are both doing yard work on a sunny morning. The neighbors are women in their early forties, each with teenage daughters enrolled as seniors in the same public high school. As they talk, Linda tells her friend that she and her husband are considering a special graduation gift for their daughter, Jen. Linda says, "Well, we just aren't sure that we should give her this gift, because it is kind of expensive, but we are pretty sure that she would like it. We talked to a friend who is a plastic surgeon, and he told me that breast enhancements are being done so often, and that technology has advanced so much that it is hardly invasive at all, and can be done so quickly. You know, Jen has always been on the small side, and she has such a pretty face. . . . We would really just like to give her every advantage that we can. We have the money, so why shouldn't we do it?"

This scenario represents an increasingly commonplace reality. According to the American Society for Aesthetic Plastic Surgery, breast enlargement surgery increased by nearly 500 percent between 1998 and 2008 among women eighteen and younger, while the increase among all age levels was 300 percent. Many doctors report that young girls come in with their parents, who are buying the procedure for a gift.[2] Aubrie, who was a seventeen year old in suburban Dallas in 2004, planned to use her parents' gift to increase her 32A cups to a C. "If my mom is offering to pay for it now, why not?," she said.[3]

Were a bioethicist to discuss a scenario like Jen's, she might try to discern the line between therapy and enhancement in order to caution against using what is still major surgery for optional cosmetic purposes. As Leon Kass points out, it is generally assumed that while it is acceptable to use technology to heal disease, it is not necessarily acceptable to use it to "augment or improve . . . native capacities and performances."[4] Few would argue that a mother taking her daughter to have cosmetic surgery to repair damage from an automobile accident, for instance, would be problematic. Since the gift of breast enhancement seems to be motivated only by the desire to become more sexually appealing, surgery could be seen as an unnecessarily risky overcompensation for that. But even with all the risks, young women today are rarely coming to the conclusion that breast enhancement surgery is at all problematic. The safer and cheaper such surgeries become, the more commonplace they will be.

Part of the reason for this cultural comfort with breast enhancement is that the distinction between therapy and enhancement is notoriously difficult to navigate. As Kass notes, the distinction easily breaks down because the definitions of enhancement and therapy are "bound up with, and absolutely dependent on, the inherently complicated idea of health and the always-controversial idea of normality." So even if one believes that technology should never be used for enhancement, one does not find a clear break between acceptable and unacceptable uses, but a continuum. What's worse, the continuum shifts according to fluctuating cultural norms. "Is it therapy to give growth hormone to a genetic dwarf, but not to a short fellow who is just unhappy to be short? And if the short are brought up to the average, the average, now having become short, will have precedent for a claim to growth hormone injections."[5] Does it change things for the parents and their decision if Jen felt, as so many young women do, that she had "abnormally small" breasts to begin with? And if the daughter wanted a breast reduction because she felt she had "abnormally large" breasts, would they be more willing to grant the surgery?

The issues become much more complicated when we consider genetic alteration of future generations, in which it is at least conceivable that all parents would choose a certain minimum breast size for their daughters, automatically putting any who are born "naturally" outside the norm and, arguably, set up to be ostracized. While we may be tempted to rely on our intuition, we quickly find that it is culturally determined. Our intuition may help us to arbitrate the difference between a woman who wants a breast reduction to appear normal and one who wants an enlargement in order to look like Angelina Jolie *today*, but it will not help us guide the parents whose daughter attends school *tomorrow* in which Angelina Jolie's breast size has become the new norm.

The difficulty in clearly separating therapy from enhancement is just one example of how ethical quandaries cannot best be resolved by appeals to what some people consider inviolable in human nature. The pervasive use of cosmetics and orthodontics proves that we are already technologically enhanced, and as enhancement surgeries become safer and cheaper, women will not be persuaded by appeals to "normal" human nature. Is there another way to approach this issue?

CHANGING THE QUESTION

Enhancement technologies, and the debates surrounding them, strike right to the core of how we define the good life.[6] In his book *Sources of the Self,* Charles Taylor argues that how individuals define the good and their relation to it is of central importance in the effort to find meaning. As Taylor emphasizes, individuals do not set the agenda for such definitions, because individuals are always situated within frameworks—what Taylor calls "constitutive questions and concerns"—that are defined by the communities of which individuals are a part. Because decisions about the good life are inherently relational (individuals in relation to each other and to the good), it "means that we understand ourselves inescapably in narrative."[7] Narrative cannot be opted out of, circumvented, or transcended.

In the scenario I described above, all the actors are seeking for the good in the way that Taylor describes; they are implicated with each other in narratives that define the good. Jen's mother and father want to give what is best to their daughter. In wanting to go ahead with the surgery, they imagine a narrative whereby the daughter's resulting increase in self-confidence gives her advantages she would not have otherwise possessed. Likewise, the neighbor wants to advise a good course of action and imagines a narrative where she offers life-giving advice to her friend. Since each one is acting out of his or her definition of the good as he or she understand it, the character and moral imaginations of each of these agents, as well as the communities of which they are a part, is essential. And so the question has now become: have these agents correctly ascertained the good? If not, how not? Why not?

In the space of all intimate relationships, such as that of mother and daughter or husband and wife, people have a variety of complicated reasons for making the decisions that they do. In his book *Better Than Well,* Carl Elliott explains that in America, the use of enhancement technologies is directly tied to the way we imagine others see us. As such, the increasing reliance on these technologies cannot be seen as simply the desire to stand out or the desire to conform, but as the outcome of either one of these desires, depending on the circumstance. In the case

of breast enhancement, one could easily imagine a woman choosing reduction or augmentation for either reason: so as not to draw attention
to herself or in order to do so. Elliott notes that Alexis de Tocqueville
had observed this dual tendency in American culture. Tocqueville "believed that social conformity and social rebellion are merely different
sides of the same coin. Both are consequences of the American preoc-
cupation with the opinions of other Americans."[8] Elliott thus views the
topic of enhancement technology as a lens through which to examine
how Americans make decisions about what constitutes the good life.
"The issues at stake in medical debates over enhancement technologies
are important, I believe, mainly because of what they can tell us about
pathologies in the way that we live. The uneasiness that many of us feel
about enhancement technologies can tell us something important about
selfhood, authenticity, and the good life."[9]

American writers have been particularly adept at pointing out some
of these "pathologies in the way that we live" and what they say about
how Americans go about defining the good life. Nathaniel Hawthorne's
story "The Birth-mark" is so compelling a look at the pathological ex-
tremes of cultural perfectionism that the President's Council on Bioethics put it first in its anthology *Being Human,* claiming that it might
best capture the dilemma of whether our flourishing depends most on
improving ourselves or on accepting our limitations.[10] At first "The
Birth-mark" seems to be merely a rather simple (and heavy-handed) account of the scientific tendency to overreach its limits. In the story,
Aylmer is an ambitious scientist who becomes obsessed with the one
imperfection in his wife, Georgiana—a birthmark on her cheek—and
accidentally kills her in the process of removing it. The story has been
used almost as much as Shelley's *Frankenstein* to proclaim that scientists
should back off from trying to play God because when they do so, they
inevitably fail.

But Hawthorne's tale, like Shelley's, is a good deal more complex
than that. While both these stories show how scientists (particularly
male scientists) end up with unintended consequences to their godlike
overreaches, they are more compelling when read as illustrations of the
character of the individuals who are tempted to use science this way.
The problem is less that a finite person is trying to play God than that

a selfish and unloving person is. Both Victor Frankenstein and Aylmer are solipsists who love their own idealistic visions more than they love actual other people. Because of these flaws, they misidentify what is actually best for the people they love. In *Frankenstein,* for example, Victor is so obsessed with his creature and his own resulting problems that he leaves his wife alone on their wedding night. It simply never occurred to him that the monster would kill her and not him.

Since enhancement decisions are always made within the context of relationships toward each other and toward the good, and those relationships are best seen within narrative, "The Birth-mark" is particularly revelatory because it is as much about motives in marriage as it is about scientific ambition. Although Aylmer claims to love his wife, he fails her miserably. "The Birth-mark" thus reveals the possibility that Jen's parents, in choosing to give breast enhancement as a gift, will fail their daughter the same way that Aylmer fails his wife: by misidentifying the good.

MARRIAGE, THE BODY, AND "THE BIRTH-MARK"

Nathaniel Hawthorne had been married less than six months when he wrote "The Birth-mark" in 1843. It seems to be, among many other things, a cautionary tale that Hawthorne wrote to warn himself about the importance of ordering his loves correctly.[11] The narrator informs us that Aylmer, the scientist protagonist so long in love with his work, had left his consuming laboratory work just long enough to persuade a beautiful woman to become his wife. Because he had so long devoted himself solely to science, he now has a problem. If Aylmer's passion for his young wife endures, the narrator insists, it will only be "by intertwining itself with his love of science and uniting the strength of the latter to his own."[12]

Aylmer soon finds that he cannot do it. Loving a real woman proves to be more difficult than loving science. Aylmer's love for his wife is a far distant third in the order of his loves, with love for himself and science taking the first two spots. Soon after they are married, Aylmer becomes bothered by the appearance of a small birthmark on Georgiana's cheek. The birthmark is in the shape of a hand. Whether the birthmark is to be seen primarily literally, as a physical imperfection, or symbolically, as a mark of flawed human nature, it is clear that Aylmer did not notice it when they were engaged. Although most married people learn quickly that one must adjust to the reality of one's partner over the envisioned fantasy, Aylmer does not make this adjustment. Very soon after they marry he wants to change his wife. He suggests, very gently at first, "has it never occurred to you that the mark upon your cheek might be removed?" (*NHT,* 119).

It had not occurred to Georgiana at all. In fact, for her whole life she thought the birthmark to be a charm, so when she discovers how much it bothers Aylmer, she tearfully confronts him for choosing to marry her. "Then why did you take me from my mother's side? You cannot love what shocks you!" (*NHT,* 119). The narrator goes on to explain that whether a person saw Georgiana's birthmark as a charm or an imperfection depended on "the difference in temperament in the beholders." In fact, many of Georgiana's previous admirers thought it to be a distinguishing mark of beauty and were "wont to say that some fairy at her birth hour had laid her tiny hand upon the infant's cheek, and left this impress there in token of the magic endowments that were to give her such sway over all hearts" (*NHT,* 119).

Since the birthmark elicits a variety of different interpretations from viewers, it is itself not the issue. The issue is Aylmer's vision, his moral imagination. Aylmer, having a scientist-idealist vision of the world, chooses to see the birthmark as an example of human finitude and death, that which he dedicates his science to overcoming. "In this manner, selecting it as the symbol of his wife's liability to sin, sorrow, decay, and death, Aylmer's sombre imagination was not long in rendering the birth-mark a frightful object, causing him more trouble and horror than ever Georgiana's beauty, whether of soul or sense, had given him delight" (*NHT,* 120). Very soon the birthmark becomes the "central point of all" in his dealings with her. Under his gaze, she is reduced to it. A story that should have been about Georgiana, or about the marriage of Aylmer and Georgiana, is now a story about a birthmark.

The fact that Hawthorne wants readers to examine the flaws in Aylmer's vision, and not in Georgiana, is clear from the effect Aylmer's gaze has on his wife. Because he is obsessed with the mark, she becomes self-conscious. Ironically, his unloving gaze brings out the birthmark even more. "Georgiana soon learned to shudder at his gaze. It needed but a glance with the peculiar expression that his face often wore to change the roses of her cheek into a deathlike paleness, amid which the Crimson Hand was brought strongly out, like a bas-relief of ruby on the whitest marble" (*NHT,* 120). Aylmer's gaze has a kind of reverse Pygmalion effect on his wife. The mythological Pygmalion was a misogynist who found women so undesirable that he made a perfect one, praying to the gods that she would come to life. Venus grants his wish, and her marble flesh warms under his hands. Aylmer is worse than Pygmalion, for he actually loves *less.* He withdraws his love from a nearly perfect woman and turns her into marble in the process.[13] He never notices that it is his gaze that makes her imperfections visible.

The image of a woman turning to marble is central to the critique that Hawthorne's story offers. Perfection, especially as reflected by classical statuary, is by definition something static or completed. What Aylmer rejects is the notion that his wife is, by virtue of being human, in process, *becoming.* This is the primary reason why his gaze must end with her death. Nothing perfect can live, for living entails the process of continually becoming something else.[14] That Hawthorne was concerned with how idealism can be fatal to the flesh-and-blood reality of marriage can be seen in a notebook entry he wrote in the early years of his own marriage, in which he blamed men who have difficulty choosing wives on the fact that they "seem as they would take none of Nature's ready-made articles, but want a woman manufactured purposely to their order."[15] With the choice of the word "manufactured," Hawthorne was saying more than he recognized.

Hawthorne's reversal of the Pygmalion story also evokes the famous New Testament injunction on how a husband ought to love his wife:

> Husbands, love your wives, as Christ loved the church and gave himself up for her, that he might sanctify her, having cleansed her by the washing of water with the word, so that he might present the church to himself in splendor, without spot or wrinkle or any such thing, that she might be holy and without blemish. In the same way

husbands should love their wives as their own bodies. He who loves his wife loves himself. For no one ever hated his own flesh, but nourishes and cherishes it, just as Christ does the church, because we are members of his body. (Ephesians 5:25–30, English Standard Version)

This passage, which would have been well known to nineteenth-century readers, slaps Aylmer in the face. Husbands are to love their wives to their perfection the way that Christ loves the church—that is, self-sacrificially. That the beloved could ever be made perfect, or "without blemish," can only be the result of self-sacrificial love. The birthmark fades whenever Georgiana blushes with love; in other words, it is only and exactly when *Aylmer* behaves properly that the birthmark is not an issue. When he transfers his attention from her to the imperfections themselves, they become even more visible, and she herself fades.

The devastating thing about the marriage in "The Birth-mark" is how fully Georgiana acquiesces to her husband's vision for her. In the course of the story, she moves from anger at his vision of her, to suspicion of his judgment, to acceptance of his vision.[16] This fact is not surprising because his critique is couched in affirmations of her beauty: she is "so nearly perfect" that it is a shame she should not be completely perfect. By the end of the story, she has so absorbed her husband's view of her that she becomes even more idealistic than he is. When he finally tells her that he can remove the birthmark (but at great danger to her), she has so internalized his vision of her that she says, "Remove it, remove it, whatever be the cost, or we shall both go mad!" (*NHT,* 128). Because she believes that Aylmer loves her, she assumes the flaw is in her expectations for herself, not in his vision. She sees his aspirations as a sign of excellence of character: "Her heart exulted, while it trembled, at his honorable love—so pure and lofty that it would accept nothing less than perfection nor miserably make itself contented with an earthlier nature than he had dreamed of. She felt how much more precious was such a sentiment than that meaner kind which would have borne with the imperfection for her sake, and have been guilty of treason to holy love by degrading its perfect idea to the level of the actual" (*NHT,* 128).

The irony in this passage is very thick. Georgiana has assented to her husband's definition of a "noble love" that has nothing to do with her actual person and everything to do with Aylmer's "highest and deepest" conceptions of it. If the possessor of the "perfect idea" of love cannot bear with a mere birthmark, then clearly he does not love at all.

It is also ironic that Georgiana becomes more idealistic *and* more realistic than Aylmer. Georgiana is the one who understands that Aylmer by his very nature can never be satisfied. She knows that the perfection he wants could only be possessed for a moment, not day to day. And so she prays that she would be able to satisfy him with such perfection, if only for a moment: "Longer than one moment she well knew it could not be; for his spirit was ever on the march, ever ascending, and each instant required something that was beyond the scope of the instant before" (*NHT*, 128). Georgiana knows that Aylmer's aspirations will find no end. She seems even to know that his aspirations cannot remain in the world as it is.

Among other things, Aylmer fails to recognize that there are various forms of transcendence, some of them good for human beings, and some not. As Martha Nussbaum argues in her essay "Transcending Humanity," it is a worthy goal to work toward "internal transcendence," which is gained, in part, by the development of virtues. Nussbaum argues that Henry James and Marcel Proust encourage this kind of transcendence by offering a "glimpse of a more compassionate, subtler, more responsive, more richly human world." Other kinds of transcendence seek instead to achieve a kind of godlike perfection. This aspiration, says Nussbaum, is the kind that should be rejected as incoherent because it is an "aspiration to leave behind altogether the constitutive conditions of our humanity, and to seek for a life that is really the life of another sort of being—as if it were a higher and better life for *us*." The ancient Greek idea of hubris, Nussbaum goes on to explain, delineates between the types. Hubris is "the failure to comprehend what sort of life one has actually got, the failure to live within its limits (which are also possibilities), the failure, being mortal, to think mortal thoughts. Correctly understood, the injunction to avoid *hubris* is not a penance or denial—it is an instruction as to where the valuable things *for us* are to be found."[17] In other words, hubris involves a misidentification of what is actually good for a human being.

For the newly married Hawthorne, the good is found in learning to love here, to love his wife for who and what she is. But Aylmer does not understand this. He eagerly returns from his lab with the concoction that he thinks is perfect and "cannot fail" to fix her. And, as to Georgiana's response, Hawthorne gives us this bizarre passage:

> "Save on your account, my dearest Aylmer," observed his wife, "I might wish to put off this birthmark of mortality by relinquishing mortality itself in preference to any other mode. Life is but a sad possession to those who have attained precisely the degree of moral advancement at which I stand. Were I weaker and blinder, it might be happiness. Were I stronger, it might be endured hopefully. But, being what I find myself, methinks I am of all mortals the most fit to die." (*NHT,* 129)

Georgiana tells her husband that she has learned that her life is a "sad possession" because she knows too much about her human imperfections to be happy and is too weak to simply endure them with hope. What the reader must remember is that Georgiana learned this perspective from Aylmer himself; she attained this "degree of moral advancement" because of his involvement in her life. She had been blissfully ignorant of her birthmark as any kind of problem before they married, and now she bemoans the lack of strength to help her to endure her imperfections. Aylmer has abdicated every bit of responsibility as a husband here, seen clearly in the fact that she would rather die than remain imperfect (*NHT,* 129).[18]

If Aylmer actually loved his wife with humility instead of hubris, that love could help her to grow into a better version of herself. Love begins with a humble acceptance of one's own finitude and brokenness, and it gently extends a helping hand to the other. In *Works of Love,* Søren Kierkegaard explains that for love to be love, it must be for the actual person one sees, not for the ideal version you want the person to be. Jesus loved Peter this way. Kierkegaard writes that Jesus "did not say, 'Peter must first change and become another person before I can love him again.' No, he said exactly the opposite, 'Peter is Peter, and I love him. My love, if anything, will help him to become another person.'"[19]

Aylmer has moved fatefully to a position where he requires his wife to change before he can love her again. Even after she declares that she is fit to die, Aylmer does not see his need to step in and to love her better. Instead he tells her: "'You are fit for heaven without tasting death! . . . But why do we speak of dying? The draught cannot fail'" (*NHT,* 129). There has been no genuine dialogue between these two people; Aylmer essentially ignores what his wife tells him that she has learned about herself by reiterating his own idealized version of her. She has never been, and is even less now, her own person.[20] In Alistair McFadyen's analysis of personhood, he reserves the word "dialogue" to mean genuine conversation between an "I" and a "Thou" wherein participants are open to each other without manipulating, or being manipulated by, the other. Drawing on Dietrich Bonhoeffer's description of the "ethical transcendence" of the other, McFadyen writes that there should be a kind of pause in a dialogue that acknowledges this transcendence. The participants need to suspend judgment in order to explore, and not immediately to oppose, the other person's viewpoint.[21] Aylmer allows for no such pause. Missing the irony that his wife understands better than he does what will become of her if she takes the potion, he gives it to her in a fit of self-confidence.

It is tempting to see the fact of Aylmer's being a scientist who develops such concoctions as merely a device for Hawthorne to make a general critique of the scientific quest for perfection. But in a story primarily about marriage, the critique sharpens into one that suggests that it is thinking scientifically and technologically *first* about solving human problems that can do the greatest damage to persons as persons within relationships. The existence of Aylmer's "concoction" here is no mere neutral device, for it is the potion's ability (or perceived ability) to solve Georgiana's "problems" in this manner that prevents (or at least delays) Aylmer from seeing his fault in the core issues in their marriage. Although his speech denies her poor self-image, his actions validate it; they say, "yes, you are the flawed one here; I will fix you." Aylmer's reliance on his quick-fix potion is not surprising; after all, it puts him in charge of solving the problem of his own dissatisfaction by misnaming it as her imperfection. Aylmer does not need to look within himself at all.

Hawthorne's contribution to the ethical questions behind the biotechnological revolution is now clear. The quick fix that technology offers deflects Aylmer from the real moral problems within him and inside his relationship with his wife. Aylmer is a moral infant. His ability (or perceived ability) to remove the birthmark and his own moral infancy mutually reinforce each other. Making a potion is easy compared to learning to love someone in spite of their human flaws, especially within the daily intimacy of marriage. Aylmer has developed none of the virtues needed for a strong marriage, and as long as he turns to technology first to fix problems, he will never develop these virtues.

And so, confident, he gives her the potion. He is so delighted to see it working that he does not even notice that she is getting paler and paler. Georgiana knows right away that she is dying, but Aylmer is so obsessed with the success of his project that he sees only the slow fading of the birthmark. The narrator tells us that the thing Aylmer wanted to dispose of was the very thing that kept Georgiana alive, that the "fatal hand had grappled with the mystery of life, and was the bond by which an angelic spirit kept itself in union with a mortal frame" (*NHT,* 130). Aylmer crossed a line, tried for a transcendence not available to mortals, and is left with nothing.

FROM BIRTHMARKS TO BREAST ENHANCEMENTS

With this dramatic and even sentimental ending, we seem to have moved a long way from the ethics of enhancement technology in the twenty-first century. Very few of our technological interventions have such dramatic consequences, and it seems that Hawthorne is simply overstating the case for the problems of scientific ambition. But Georgiana's physical death is not the issue here. Her death is only the tragic exclamation point on the various ways that Aylmer already killed the woman he claimed to love. His love should have brought her more into life, more into herself, more into the good for her; instead, it drained her of her distinguishing features.

To return to the graduation gift scenario, there are some obvious and important differences. Parents, of course, have a different relationship to their daughter than does a husband to a wife. Also, Jen's parents are not obsessed with transcendence per se; they are just trying to do what they think is best for their daughter. Aylmer's fixation on his wife's birthmark was, arguably, more about it as a symbol of mortality or human imperfection in general than it was about her appearance.

But the continuity between the story and the scenario is stronger than the differences. The main point of comparison is that both the parents and the husband claim to be motivated by the desire to give the beloved what is best for her. Both desire to give the other a gift that is connected to a narrative that defines the good life for the recipient. Both examples, therefore, force the question of the character of the gift giver, as well as the question of the good entailed in the gift itself. Do the givers correctly discern the good for the persons they love? Do they give good gifts?

First, notice that both Aylmer and Jen's parents exercise a great deal of power in choosing what they give—in the offering of the gift itself. Both imply that what they give is good, that it is something that will only benefit, and not harm, the recipient. In intimate relationships, the giving and receiving of gifts is not equivalent to the person going out to acquire the thing themselves. In Jen's case, the mother, Linda, may be only giving Jen something that she *thinks* Jen wants. Perhaps, for instance, she overheard her daughter make a comment on her breast size. But even if Jen was dissatisfied with her breasts, it does not follow that she necessarily wants augmentation, and in offering the enhancement surgery, Linda would be encouraging a wholly different way for Jen to think about her body and how to "fix" it. If Linda had overhead Jen saying she wanted the surgery, or that Jen had been the one to request it, the offer of the gift still works the same way: it validates the daughter's explicit or implicit desire to be more attractive according to current standards of female attractiveness. The offer also stands as an agreement with Jen's self-assessment that "I have small breasts; that feature can and should be improved." It also necessarily relocates everyone's view of Jen away from her as a whole person and onto one part of her body. A narrative that should have been about Jen's achievements as a young woman who graduated from high school and is commencing the next stage of her life becomes a narrative (at least somewhat) about her body.

Furthermore, both Aylmer and Jen's parents give gifts that imply a rejection of the *givens* for both Georgiana and Jen. In Wendell Berry's reflection on marriage, he notes that marriage rests on unchangeable givens of "words, bodies, characters, histories, places" and that there is relief and freedom in knowing that these givens are as real as the earth on which we walk.[22] Part of the reason why acceptance of these givens is so important in any relationship, but particularly within a marriage relationship, is that it reminds the parties that no person is completely free to will the shape of his or her life.[23] In denying the givens of the human body (its limitations, however conceived) Aylmer and Georgiana, and potentially Jen and Jen's parents, choose a route that encourages the denial of negative givens as a way of life. Such a denial, insists Berry, is, somewhat paradoxically, to give up on hope for the good, for "to forsake the way is to forsake the possibility. To give up the form is to abandon hope."[24] Another way to put this might be to say that with limitless possibilities comes the despair and alienation of a never-attainable ideal.

Since the offer of a gift of breast enhancement comes from a particular view of Jen's "problem" and of the good life, it is no different in kind from Aylmer's assessment of the birthmark as something that makes Georgiana ugly and keeps her from perfection. As such, Hawthorne's story can prophetically predict how such an assessment by a lover (Jen's parents) will be received by the beloved (Jen). As in Georgiana's case, once the problem is named a problem, and the gift of a solution is offered, the recipient of the gift is effectively cornered, no matter what she ends up believing about the motives of the giver. While Jen can accept or reject the solution, or even the values behind it, she cannot reject the fact of their assessment of the problem. In any relationship, that assessment will have a lot of power, but it particularly has power in the asymmetrical relationships represented in Hawthorne's tale and in the enhancement scenario. The parents, like Aylmer, are the ones with the real choice; their vision is the vision with the most power in Jen's life in this particular scenario. By encouraging the daughter to have the surgery, and even more by paying for it, the parents lessen her freedom to choose not to have the surgery. They also lessen her freedom to choose to insist that she be valued according to a different standard from physical appearance.

These facts highlight the truth that because we live in community, very few decisions we make, particularly in the area of human enhancement, are made in true freedom. Georgiana's freedom to choose has been altered by her husband's decision to see the birthmark as an unacceptable symbol of human imperfection and mortality. The daughter's desire to change her body is a direct reflection of her acceptance (with help from the parents) of the gaze of society as authoritative in deciding what is beautiful and what constitutes the good life. In each case, interpretation of the "imperfection" has been closed down quickly and accepted with little argument. Part of what is being rejected is the difficulty of interpretation; to some degree, Aylmer and Jen simply follow easy, culturally given answers to the question, what is the body for?[25] In Jen's case, the interpretation ("I have small breasts; I should get surgery") is the easiest one to make because she has assessed society correctly: she will have more power of a certain sort, because Western culture does afford greater respect to women based on accepted norms of physical attractiveness.

Even Greta Van Susteren, the TV reporter who was known for being tough, smart, and to the point, succumbed to these values and got a face lift. Although she claimed that she acted freely and without any external pressure, is that really the case? No less than Georgiana, Greta Van Susteren embraced the narrative of the good life as defined and constructed by her viewers. Her decision made it that much harder for other women in the industry (or any other women watching her) to decide against surgery, or to decide in favor of a different narrative of the good. She had represented the possibility that an average-looking woman could make it, even in television, with intelligence and personality. As Susan Bordo put it, "when Greta had her face lifted, another source of inspiration and hope bit the dust."[26] The more pervasive that enhancement technologies become, the harder it will be to imagine alternative narratives of the good, precisely because no one else is modeling those alternatives.

One could make the argument that breast augmentation is an en-hancement that is no different in kind from orthodontic work. I ac-knowledge that it is difficult to discern the difference, particularly without resulting to the language of therapy verses enhancement. Very few parents discuss the issue of orthodontic work with their children; if parents can afford it, they get the work done for them. But there are some important differences. One is that having straight teeth is already the norm in America. Parents have already lost their freedom of choice; leaving their children out would be a decided disadvantage to them and, one could argue, would be an unloving act. Breast size, unlike crooked or straight teeth, is still a shifting fashion target, and surgery remains the exception, not the rule. Therefore, right now parents who offer their daughters such alterations are making a different statement and, I would argue, are acting out of corrupted definitions of the good. Their definition of where the good life is to be found—in something that their daughters do not naturally possess, but must transcend their own bodies, in some way, to attain—is the true gift that the daughters are receiving.[27]

Because of the wide range of motives behind human behavior, it is unwise to make blanket statements about the application of any en-hancement technology. The better course of action is to encourage each person who is making decisions regarding any available and emerging enhancement technologies to fully investigate their vision of the good, to query the source of that vision, and to imagine alternative narra-tives.[28] Aylmer turned to technology as a quick fix, and this turn kept him in moral infancy. In the case of Jen and her parents, the choice of a technological quick fix would be no less devastating. When the issue of her body image emerges, the parents could use the opportunity to lov-ingly affirm their daughter and to challenge society's claim on her ulti-mate source of value. They could invite her into a different narrative, a different answer to the question, what is the body for? But if they choose to give the gift of enhancement, they may be consigning her to a life in which she constantly wonders whether she can ever be good enough. Hoping to give a good gift, they may be giving something that they never intended to give: self-contempt.

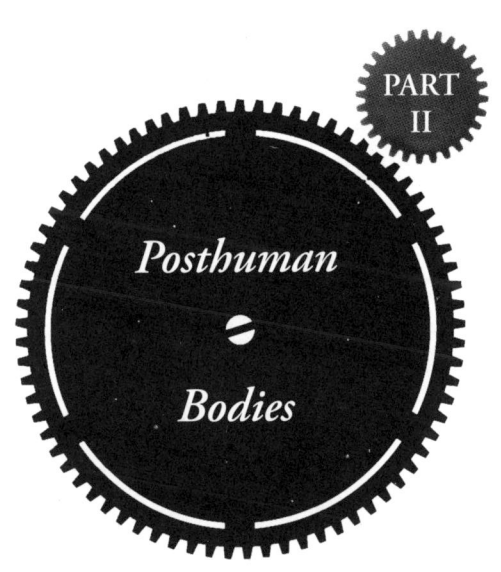

PART
II

Posthuman

Bodies

CHAPTER 3

The Faces of Others

George Saunders, James Tiptree Jr., and the Body for Sale

> *I always think that art should comfort the oppressed and oppress*
> *the comfortable.*
>
> —George Saunders

In April 2009, a video clip posted to YouTube generated over thirty million hits in one week. The clip featured Susan Boyle, a previously unknown singer, performing on *Britain's Got Talent,* a reality TV show similar to *American Idol.* But the clip generated interest not because Boyle went on to be crowned the winner of the show, but because she most likely would not be. Unsuspecting viewers watched as this homely, overweight, forty-eight-year-old spinster walked on to the stage and surprised everyone by a spectacular performance of "I Dreamed a Dream." As she began to sing, the camera moved backstage, where one of the hosts of the show looked into it and said, "I betcha didn't expect that, didya?" After Boyle finished singing, one of the judges gushed praise and apologized: "Everyone was against you."

Boyle's video crisscrossed the web through e-mails forwarded by people championing her as their new hero. People said they were inspired, even moved to tears, by it. But why? Certainly some viewers felt that they were watching the humbling of both the elite judges and the prejudiced audience; Boyle's confident self-possession felt to them like the show-biz equivalent of someone "sticking it to the man," even as they acknowledged their own participation in prejudging her. They saw her as being able, finally, to live the dream she was singing about.

But in spite of what viewers wanted the performance to mean, their response was the one scripted by the writers of the show. It did nothing to challenge the show's basic persuasion that only the beautiful and talented should and will rise to the top. *Britain's Got Talent,* like *American Idol,* thrives on the expectation that viewers will unmercifully judge, idolize, and reward the godlike celebrities among us. When the host said, "I betcha didn't expect that," it merely completed the show's framing of Susan Boyle's face as ugly. It confirmed that the audience did not, indeed, expect that an ugly woman would appear on the show unless it was to mock her. Though she was complicit, the producers used Susan Boyle. They counted on the fact that after a momentary twinge of guilt, the audience could revel in its own magnanimity in being able to recognize the talent of the ugly duckling, all the while never being challenged to think of her as *anything but* an ugly duckling. At the end, most viewers were probably thinking, "Wow, it's a shame that she's so ugly, because, man, can she sing!"[1]

The video clip reveals many things about contemporary culture, including the generation and propagation of standards of beauty, the role of mass media in society, consumption of the other in a capitalist system, and the cult of celebrity worship. But I bring it up to illustrate what fiction writers have always known: that framing matters. The network framed the viewers' experience of Susan Boyle from beginning to end, largely by the types of questions the viewers were permitted to ask or be asked. Statements like the host's "I betcha didn't expect that!" may challenge the viewers' expectations, but they do not reframe the issue. The frame holds; the moral imagination of the viewer has not been challenged. No one saw Susan Boyle any differently from how the network calculated that she would be seen.

Part of what separates popular culture from the work of literary artists is that the literary artist, because she is concerned with the limits of vision, often works by deliberate *reframing*. Speculative fiction—a larger category in which I include most types of science fiction—is a particularly interesting example of this work of reframing because the genre is so readily dismissed by the literati as mere pop culture.[2] But many of the best-known writers of speculative fiction do something that only they can do: they rely on the conventions of the genre even as they challenge their typical readers' most cherished assumptions. Indeed, some science fiction writers deliberately subvert the very paradigm on which their stories depend: in their case, the general ascendancy of science and technology, and the way that ascendancy affects the way individuals see (that is, frame) one another. So-called cyberpunk fiction, such as William Gibson's *Neuromancer*, is probably the best example of this subversion. The embodied, particular, and personal experiences of characters in these novels push against a highly technologized world that seems to be committed to reducing individuals to information patterns.

This chapter describes two American writers who subvert the framing power of technoscience from different traditions in fiction. George Saunders writes satire, and James Tiptree Jr. writes cyberpunk. Both are concerned not only with the framing power of technoscience but also with the way that it operates and uniquely thrives within the larger frame of consumer capitalism. These writers believe that technique produces and shapes the "economic man it needed"[3] and that consumer capitalism gives technique its special power. Specifically, the combination of the rule of technique and the reign of consumer capitalism leads to a defacing commodification of the person.[4] Thus, both writers make an effort to reveal the power of technoscience to frame experience by heightening it in some way (often by the use of the grotesque), but they do so in order to suggest an alternative. Their narratives strive to recover the human person as that being who must be fully faced with responsibility, not defaced by market-driven relationships.

HEY MOM, LOOK AT ME! I CAN SPEAK!

Although George Saunders has a biting satirical style all his own, he did take a hint from one of the undisputed masters of satire, Jonathan

Swift. And the hint is this: if you really want to make an impression on your audience, use babies. Or more to the point, use babies to show that "do not use babies" is one maxim still left sacred (for the most part) in a world given over to utilitarian values. And so Saunders's opening story in his brilliant collection *In Persuasion Nation,* "I Can Speak!™," features a baby who is being used.

The style of *In Persuasion Nation* insists that in a nation benumbed by advertising campaigns and the glamour of technology, the best way to wake people up is by extreme satire. Satire often draws on the grotesque, an artistic method whereby two incongruous things are drawn together to make them stand out. Saunders follows the lead of Flannery O'Connor, who famously defended the grotesque by arguing that "you have to make your vision apparent by shock—to the hard of hearing you shout, and for the almost blind you draw large and startling figures."[5] Saunders believes that Americans have bought into the glamour of technology so deeply that we will buy anything promised to improve our lives, even when that "improvement" is absurd. We've sold not only our souls to the technoscientific promises of consumer capitalism, but those of our children, too, whom we effectively deface when we view them as commodities for our manipulation and pleasure. So Saunders unmasks this paradigm—with a mask.

"I Can Speak!™" is a letter written by a desperate salesman, Sminks, to a customer, Mrs. Faniglia, to try to get her to reconsider her decision to return an item she purchased from KidLuv, a company supposedly dedicated to providing "innovative and essential development tools for families."[6] She had apparently returned the I Can Speak!™ device (ICS), a mask designed for babies to wear so that their inarticulate speech can be artificially translated into witty, preprogrammed, and personalized expressions. Sminks offers Mrs. Faniglia an upgrade and tries to convince her of how wonderful it is to be entertained by a "much, much cleverer" baby. "For instance," Sminks reminds her, "you might choose to have [your son Derek] say, on his birthday, for example, 'MOMMY AND DADDY, REMEMBER THAT TIME YOU CONCEIVED ME IN ARUBA?'" And if the dog comes to give the baby a lick, "you could make Derek say (if your dog's name is Queenie, which our dog's name is Queenie): 'QUEENIE, GIVE IT A REST!' Which, you know what? *Makes you love him more.* Because suddenly he is articulate" (*PN*, 6).

The first thing to notice about Saunders's story is that it is not a traditional story but a letter. This one-way communication is symbolic of the "persuasion nation" of infomercials and advertising in which rhetoric is reduced to the effort to persuade consumers that they need something they would otherwise not even think to want. Mrs. Faniglia, the recipient of the letter, is faceless and nearly voiceless in the story, except for the one voice the consumer is always allowed: her desires as expressed by her purchasing power and choices.

The power of Saunders's satire comes from its exaggeration of patterns and habits normally invisible to most Americans. The story is a perfect example of Albert Borgmann's point in *Technology and the Character of Contemporary Life*: that the more firmly established a pattern is, the more invisible it becomes in society, until finally it requires extreme effort to reveal it. "Living in an advanced industrial country, one is always and already implicated in technology and so profoundly and extensively that one's involvement normally remains implicit. . . . Technology is the rule today in constituting the inconspicuous pattern by which we normally orient ourselves."[7] The exaggeration of the satire serves to make the inconspicuous pattern conspicuous; it reveals that which had been hidden. It does this by mocking how consumers neither question nor understand the fact that they turn to technology to solve problems that had been invented by corporations primarily to create demand for a technological product that would solve those problems.[8] "I Can Speak!™" makes us laugh and cringe precisely because no one within its world is laughing or cringing at the idea that it is somehow a problem now that a baby cannot say anything but "glub glub glub." Mrs. Faniglia never doubts that having a baby speak would be desirable; after all, she did purchase the device to begin with. She returns the product because of its performance, notably, that it "takes on a 'stressed-out look when talking that is not what a real baby's talking face appears like but is more like some nervous middle-aged woman'" (*PN*, 5). This is why Sminks attempts to solve her concerns by offering her an upgrade to the ICS 2100, with which "your baby *looks just like your baby*" (*PN*, 5; emphasis in original).

By having Sminks attempt to cajole the woman with an upgrade, Saunders reveals the ludicrous extent to which consumers have been given over to what Borgmann calls the "persistent glamour of the promise of technology."[9] Similarly, Jacques Ellul argues that one of the hallmarks of a society given over to the rule of technique is an inability to judge the real value of the ends that the technology will serve. What excites the crowd instead is performance, and technique is its instrument. "What is important is to go higher and faster; the object of the performance means little. The act is sufficient unto itself."[10] Who cares if the I Can Speak! device does the talking instead of your child—just look at what it can do! As Sminks writes, "there is something great about having your kid say something witty and self-possessed years before he or she would actually in reality be able to say something witty or self-possessed" (*PN,* 7). The imperial vagueness of that "something great" is what Saunders is trying to expose.

Since the ICS device is designed to enhance children, it raises a whole other set of questions about the use of technology to enhance people who literally have no say in the matter. In *Better Than Well,* Carl Elliott explains how Americans are particularly susceptible to the promises of technology when driven by the desire to give the best advantages to their children. What nearly all of the debates over enhancement technology for children have in common is the issue of self-esteem; parents who are concerned for their child's self-image quickly turn to technology to gain advantages in social, intellectual, and athletic arenas.[11] This concern is what Sminks plays on right away by arguing that the ICS device "offers rare early-development opportunity for babies and toddlers alike" and that he, for one, is not "going to take any chances" with his own son not having the product. Saunders pokes fun at corporations like Baby Einstein that invent ways to help parents worry about their child's self-esteem and development far before they become an issue for the child. The companies convince parents that they need technology to stay on the cutting edge of learning.

The following is a quotation from a press release for VTech, a real-life company that develops technology for children:

To meet kids' growing need for innovative technology, VTech®, creator of the Electronic Learning products category and a leader in merging technology, education and fun, today unveiled its 2010 product line at the International Toy Fair in New York City. The most talked about and innovative toys coming to market this year include MobiGo™, a touch-screen handheld game console; FLiP™, the world's first children's animated e-reader; and additions to the already popular infant line including VTech's first-ever electronic bath toys. Parents can trust these new cutting-edge products to aid in their children's development while showcasing how learning can be fun.[12]

Sminks similarly tries to appeal to an infant's supposed "growing need for innovative technology" when he writes that he feels that "a baby, sitting in its diaper, looking around at the world, thinks to itself, albeit in some crude nonverbal way: What the heck is wrong with me, why am I the only one saying glub glub glub while all these other folks are talking in whole complete sentences?" (*PN*, 8). Saunders exaggerates what nearly all enhancement technologies have in common: they are viewed as quick and painless routes to healthy self-esteem. Because he has the ICS device, when Billy "hears a competent, intelligent voice issuing from the area near his mouth, that makes him feel excellent about himself," which, Sminks says, makes him feel excellent about Billy, too (*PN*, 9). The promise of technology is complete. Never mind that the product will certainly delay and possibly prevent the child's actual speech development.

This rank appeal to parental concerns highlights the central irony of "I Can Speak™": the fact that the product does not help the baby at all. The story's satiric energy works so well because the device literally defaces the baby and renders him voiceless. The critique works on a number of levels. First, the ICS mask is a grotesque symbol of the way that, through the power of advertising, all commodities become pure, mirrorlike surfaces, merely reflecting our desires to become something else. Borgmann points to research done by Stephen Kline and William

Leiss, who conclude that "the mask of the fetishized commodity, having incorporated the abstract qualities of promised human satisfaction, has more recently still become mirrorlike, reflecting back the vague and distorted images of well-being to be achieved in consumption."[13] In other words, the ICS device turns the baby into the commodity that is merely a reflection of some vague desire to be the parent of a superbaby. The baby's actual face, representative of the baby as a discrete person, recedes farther and farther from view.

Second, Saunders's satire exposes how easily parents who live in "persuasion nation" fall prey to the notion that children exist primarily for the parents' own self-actualization or to help them achieve even loftier social goals. As Amy Laura Hall has recently illustrated, the American eugenics movement capitalized on parents' desire to forge a "new worldwide domestic order" by promoting "properly calibrated, usefully capable children."[14] Consumer culture wedded to technoscience gives parents the sense that they control the outcome of their children. In "I Can Speak!™," the parents convert what is typically "useless," the baby's speech, into something they can use for their own satisfaction and delight. The salesman tells Mrs. Faniglia that his own son Billy is "much, much cleverer" with the upgrade than he was with the original ICS, and he says such "wonderful things . . . and is not nearly so, you know, boring as when we just had the ICS1900" (*PN,* 9). For example, when Billy wants to go outside and his pants aren't on yet, "he'll say: 'HOW ABOUT SLAPPING ON MY ROMPERS SO I CAN GET ON WITH MY DAY!'" (*PN,* 8), a text written by Billy's father for the parents' amusement. Sminks's belief in the device is so apparently strong (or so he wants Mrs. Faniglia to believe) that he encourages her to ignore her baby's cues that he doesn't want to wear the device—he is "sort of flinching" when it speaks for him—by slowly breaking him in by increasing the product's use by ten minutes each day. The salesman has trained his own son so that the child wears his mask even while sleeping, which means, effectively, that the product has entirely eclipsed Billy's natural appearance in the world. He now has no face at all.

In an interview, George Saunders explained that art, for him, was to pay attention and that paying attention is love. "My particular kind of attention is satirical. I love America, but I'm suspicious of it: it's the fat kid with all the toys and needs to be looked at honestly."[15] In the story "I Can Speak!™" it is the fact that parents so easily surrender to the disappearance of the faces and voices of their own children that encourages this sort of honest look. Saunders's satire reveals how the commodification of enhancement technologies can compromise our ability to love others, even those closest to us. The goal of all technology—the glamour of its promise—is to make people's lives easier. But the hard fact that most parents need to learn is that not everything *should* be made easier for children. Gilbert Meilaender argues that there is a vast difference between attaining a particular result in our children and helping them to become persons of character. Referring to the recent tendency to medicate children to control behavior, he also reflects that "perhaps a clinician could medicate them and they might feel they had been 'helped,' but we should not be entirely indifferent to the distinction between help that works 'with' them as they struggle to become persons of a certain sort and help that merely works 'on' them, making adjustments of biochemistry."[16] He goes on to argue that the very restlessness that the medication is designed to reduce might be something a child needs in order to direct himself toward the good life. In Saunders's story, the company ironically called KidLuv necessarily thins out parental love because it promises that the product will make "you love the child even more," but it achieves this effect through changing the child, not the parent (*PN,* 7). These parents, in short, will never learn to parent. Saunders is relentless on this point: the satire is at its peak when Sminks tells the customer that "we at KidLuv really love what kids are, Mrs. Faniglia, which is why we want them to become something better as soon as possible" (*PN,* 10).

Children are still the biggest challenge to an individual's self-absorption. Parents, even those who adopt, don't really get to choose their children; they just suddenly find themselves living at home with complete strangers they must learn to love. While it may be easier to love an other who is not an other at all, it can no longer really be called love. This is why Saunders's satire is focused on the face. The face represents personal uniqueness in a way that nothing else can, which is why Robert Spaemann, Emmanuel Levinas, and Richard Ford (just to name a few recent scholars who discuss this issue) insist that love for the other begins with the true recognition of the *face* of the other. Although Levinas's philosophical commitments draw him more toward the idea of the face as symbolic of human vulnerability than personal uniqueness per se, in his early work he singles out the face as the locus of speech: a reminder of the otherness of the other. "The face, preeminently expression, formulates the first word: the signifier arising at the thrust of his sign, as eyes that look at you."[17] Levinas explains that this otherness, represented by the face, is a kind of silent language of exteriority of a living presence who is not reducible to my I. So in taking over the face and the speech of the baby, the ICS device doubly silences the voice of the other. It is the I conquering, not loving, the Thou. As Spaemann has argued, persons exist only in the plural; for another to be a person, there must be a "letting-be" that resists the "tendency of all living things to overpower others."[18] KidLuv thus eliminates the possibility of the very thing the company promises. Consumer capitalism, thriving best on dissatisfied self-absorption, wouldn't want it any other way.

Careful readers of my argument here might object that this satire merely exaggerates our tendency to impose our own wills and desires on our children and that this failing has nothing to do with technology per se. Ronald Green would be one such reader, and his book *Babies by Design* is exemplary in its careful evaluation of the ethical problems raised by the potential for germline genetic interventions. Although germline genetic interventions (the alteration of an individual's genotype to select traits) are not yet possible, they are the goal of advocates of reprogenetics. Green, one such advocate, takes on what he sees as the four main classes of objections to such interventions, including the objection that we might love our children less if they do not live up to the lives that we had planned for them genetically. To this reasoning he applies what he calls the PLAAP principle: parental love almost always prevails. And so he argues, "parents will try to produce the children they desire, but in almost all cases, they will love the children they get no matter what qualities they possess. Parents who wish for abled children will accept and love disabled ones; parents who try to have enhanced children will accept and love average or disabled ones. Those who think otherwise and believe that gene interventions imperil children do so because they mistakenly apply the culture of consumerism and manufacture to the parent-child relationship. In doing so, they overlook the power of the PLAAP principle."[19]

I like the idea of the PLAAP principle and generally agree that parental love will prevail. But to assume that consumer culture does not significantly affect the way parents think about their own children is naïve. It is true that putting a mask on a baby to make it speak prematurely is not equivalent to a parent being able to choose certain genetic characteristics (for example, height, weight, musical or athletic ability). But the purpose of satirical exaggeration is to reveal attitudes to which we may be oblivious. What Green misses in his analysis, and what Saunders exploits with his satire, is the unique way that enhancement technology encourages the worst in parents, the way that it draws parents further and further away from viewing the lives of all other people, including their own children, as a gift received rather than as a choice they made. In other words, one could argue that far from "mistakenly applying the culture of consumerism and manufacture to the parent-child relationship," we have not even begun to see how the consumer culture has already changed that relationship. Americans have grown so accustomed to viewing others through a consumer's lens that they do not question the impact of enhancement technologies that are becoming increasingly more available, more powerful, and more permanent.

Green ends this part of his argument by asking readers to consider the fact that while parents used to leave everything to chance about having children, they now use contraception to plan the size and timing of their families. "Has any of this impaired parenting?" he asks, and "do we reject our children more often than people in my grandmother's generation did?"[20] These are good questions, but they are also characteristic of a certain kind of historical myopia. The question of exactly how all reproduction technologies have changed American culture's view of children as a whole is quite a bit more open than this dismissal suggests. For example, what happens to potential parents when they have been led by technology to expect increasing amounts of control over the lives of their children? Will parents view children more like outcomes? What happens to their view of other people's children in that scenario? Until these questions are considered, it makes sense to urge extreme caution in the adoption of something as powerful as germline genetic engineering. Saunders's story reveals why.

THE REAL HAIRY THING

The rise of reality television has seen the birth of a collection of shows that trade on the promise of technology to give mere mortals the lives of the gods. In the early years of this century, ABC's *Extreme Makeover*, Fox's *The Swan,* and MTV's *I Want a Famous Face* all recruited contestants who believed that a glamorous life was a few surgeries away. In each of these shows, contestants voluntarily submitted to radical plastic surgery and other regimes of self-improvement, and they agreed to do so on national television. The women who participated in *The Swan* even agreed to compete with all of the other transformed women in a final beauty pageant. The result was a spectacular display of the "narrative strategy of 'wish fulfillment.'"[21]

While it is tempting to believe that whatever Fox does doesn't matter, the facts speak otherwise. The humanities more fully caught up with this concern when the journal *Configurations* devoted a two-part volume to essays that address the cultural significance of these shows. Taken collectively, the essays in *Configurations* argue that these shows do more than simply reflect consumer desires for an escape via technology into another life; they also shape and feed it. In her introduction, Bernadette Wegenstein explains that the "data speak volumes," notably, that the peak year of the existence of these shows, 2003, saw a 44 percent increase in surgical and nonsurgical enhancement procedures. After the shows tapered off, the increases tapered off, too; and the numbers remain constant at 11.5 million procedures per year, an increase of 446 percent since 1997.[22] Clearly, this narrative of the life-changing possibilities of technoscience drew millions of female viewers into its paradigm like moths to a flame.[23]

James Tiptree Jr. was a writer accustomed to offering counternarratives. Born Alice B. Sheldon, Tiptree hid behind her nom de plume for almost ten years, writing "fiercely, blazingly feminist" short stories in a genre dominated by men.²⁴ While shows like *The Swan* unblinkingly tell audiences that they are lost in their ugliness and in need of a technological fix to give their lives value, Tiptree insists on pointing out the self-contempt behind their promise. Forty years ahead of its time, Tiptree's story "The Girl Who Was Plugged In" prophetically challenges the facile cultural assumptions about the body, desire, love, and fulfillment that fuel these television shows.

The story opens by insisting that the readers look at P. Burke, one of "the ugly of the world" who is about to commit suicide to escape the miserable distance between herself and the lives of the "gods"—the celebrity others—whom she worships. But before she is able to kill herself, P. Burke is approached by GTX, a corporation that specializes in surreptitiously creating and placing beautiful people where they can be seen by millions of viewers using certain products. GTX offers Burke a new life, but she must submit to putting her real body five hundred feet below ground, attached electronically to a "waldo" named Delphi, a beautiful body (with limited sensory capability) that her brain impulses would animate and live through. Burke agrees and goes on to enjoy the life she thinks she wanted, being watched by millions of adoring fans. But then she falls in love with Paul, the son of a GTX executive who does not know the truth about her. Ashamed of her real person that lives underground, Burke nevertheless believes that Paul can truly love her, that she really is Delphi. Paul eventually discovers that something is not right, but he believes that Delphi is a real person who is being controlled by someone else, so he goes to unplug her, only to discover the real P. Burke and reject her, violently pulling out wires until both Burke and the waldo Delphi die.

Needless to say, this is not a hopeful story. But its exaggerations are poignant and revelatory for a culture in love with technological solutions. First, "The Girl Who Was Plugged In" suggests that technoscience is more than a neutral shell; it is itself a narrative force that provides structure in the life of P. Burke just as it does in the lives of contestants and viewers of *The Swan*. Both the show and the story do not suggest or look for other options for these women; instead, they illustrate a reality conditioned by expectations that there is a technology available to improve anything in a woman's life that seems inadequate. They lend support to Albert Borgmann's assertion that liberal democracy always answers the question of the good life along technological lines.[25]

Borgmann's analysis draws a great deal from Heidegger, and, indeed, his abstract but provocative essay "The Question Concerning Technology" provides some insights here. Heidegger's primary concern is to discern the essence of technology, which he describes in terms of enframing, *gestell*. The problem for Heidegger is that technology frames the world in a limiting and pernicious way, specifically by starting from the premise that everything is under human control. Modern science and technology (as best seen, for Heidegger, in physics) "sets nature up to exhibit itself as a coherence of forces calculable in advance, it orders its experiments precisely for the purpose of asking whether and how nature reports itself when set up in this way."[26] In other words, since technoscience assumes that nature exists solely for the forces it can produce, it does not and cannot reveal the world outside of the frame of these types of questions, outside of its experimental paradigm. Consider the Rhine River. Under the technoscientific paradigm, according to Heidegger, the river can only be seen for what it produces for us (basic economic utilitarianism), and loses its "original revealing." For Heidegger, we thereby lose the opportunity to "experience the call of a more primal truth," which constitutes a loss.

Heidegger's concern is phenomenological; he is focused on how the world comes into view. Television shows like *The Swan* depend on controlling exactly how the worlds of these women come into view for the audience. Before their transformation, the women are dressed in drab gray underwear and framed on screen with graphics that rotate their bodies, highlighting and targeting areas that need to be fixed. The experts and the audience look on, and there is no room to disagree with the camera's assessment. It is also assumed that the women agree with the various assessments of their bodies made by the experts.

Technologies like video editing do not cause this kind of dehumanization, but they do make it easier. Dehumanization depends on the exertion of control over how the subject is viewed. *The Swan* is completely in earnest; there is no irony and no room for the women to tell their stories beyond what the editor has arranged for the audience to see; in short, the editor uses them.[27] This is nothing new, of course, but it points to the reason why fiction must be called on to respond to these spectacles. Since narrative is what is curtailed when inhumane treatment is applied, narrative is what is required to restore dignity. This is why Martha Nussbaum insists that a complete picture of justice requires sensibilities in us that only fiction can provide. In her book *Poetic Justice* she argues that "very often in today's political life we lack the capacity to see one another as fully human, as more than 'dreams or dots.' Often, too, those refusals of sympathy are aided and abetted by an excessive reliance on technical ways of modeling human behavior, especially those that derive from economic utilitarianism."[28] It is that lost capacity, argues Nussbaum, that makes sensitivity to story an essential ingredient in public discourse. Storytelling can recover the particular faces that may be lost in the utilitarian way of seeing the world.

Fiction saves faces by reframing them, by forcing readers to regard them in a different way from how they want to. This is why it is important to note that the backstories of the contestants on *The Swan* are all similar to P. Burke's story. The women considered themselves the ugly ducklings of the world, and each of them believed that she would have a new life if she had a new body. Each of them had also been the victim of some kind of dramatic abuse or loss, just as P. Burke had been raped at age twelve by some drunk "freak lovers." And perhaps most interesting of all, in *The Swan* the women who ended up winning the final contest were, as Bernadette Wegenstein learned when interviewing the producer, those who most thoroughly surrendered to the beauty culture, those who "gave in" the best.[29] In other words, those contestants who were most convinced that technology could solve their problems were rewarded for submitting in a way not substantially different from being hooked up to the Delphi waldo. The difference is that Tiptree's story, through the grotesque exaggeration enabled by the conventions of science fiction, reframes the exchange to reveal the transformation for what it is: acquiescence to the larger cultural narrative that only the beautiful people matter. Like all good science fiction, "The Girl Who Was Plugged In" does much more than reveal the power of the technoscientific paradigm. It also reveals the dependency of that paradigm on corporate culture's manipulation of ordinary human desire. In other words, Tiptree is prophetic; she is an aggressive truth teller. In Walter

Brueggemann's language, she critiques the dominant consciousness. What she is shouting is that capitalism preys on people's displaced desires for the lives of others, and technology is its most irresistible bait.

As if to emphasize this prophetic goal, the narrator of the story, a young woman, addresses the hearer (and readers) quite violently: "Listen, Zombie. Believe me. What I could tell you—you with your silly hands leaking sweat on your growth-stocks portfolio."[30] The hearer appears to be someone who cares only about making money on investments. The narrator doesn't pull her punches: "You doubleknit dummy, how I'd love to show you something" ("Plugged In," 43). As she proceeds, the narrator repeatedly commands the hearer to not be distracted by the technology in the city of the future, but to look at the "rotten girl" that is at the center of her story, even when they do not want to. Prophetic energy is already evident in this insistence, for it suggests that not everyone wins in the utopian city of the future, and that the losers need the artist's attention most of all. Listen, she seems to insist, for I am going to tell you an important parable.

Tiptree further emphasizes the parabolic and prophetic nature of her tale by giving the girl the name P. Burke, where the *P* stands for Philadelphia. When Burke decides to become plugged in, her waldo goes by the name Delphi. Delphi, of course, was the seat of the famous Greek oracle visited by Oedipus when he learned of his ultimate fate, a fate that would be precipitated by illicit desire. The oracle's answers were delivered by a priestess who went into a trance, and the answers were apparently incontrovertible. So the name Delphi/Philadelphia is a fusion of the idea of an oracle with a name that means "the city of brotherly love." The story itself becomes that fusion: it is an oracular pronouncement on the ultimate fate of love and desire in an American city.

The genius of Tiptree's story is in its revelation of the true nature of the forces behind consumer capitalism and its penchant for technological solutions. The thing that drives consumer capitalism and the biotechnological revolution is more than materialism or greed. It is a basic desire to be loved, a desire that is displaced onto the objects that those who appear to be the most loved—celebrities—possess. Tiptree emphasizes this fact by revealing that in this future city, the "Huckster Act" laws forbid advertising, so companies sell everything only by product placement. They put as many of their products into as many of the beautiful people's hands as possible. P. Burke is not alone in her desire to be loved and in her admiration of the beautiful people. The narrator emphasizes that the "whole boiling megacity, this whole fun future world loves its gods" and that Burke is just one among many who has her "soul yearning out of her eyeballs" ("Plugged In," 43).

This story thus exaggerates how the American economy exploits what René Girard calls "mimetic desire."[31] In *Deceit, Desire, and the Novel,* Girard argues that the novelistic imagination is that which can identify that the self does not generate its own desires but follows an identifiable structure of imitation. Girard gives the example of *Madame Bovary* to describe how the structure is triangular: Emma Bovary desires to be like the heroines in the books that she admires, so she desires objects that she associates with those heroines. Her desires are not her own. In Tiptree's story, P. Burke is not exceptional, only desperate; she goes further in her attempt to attain possession of the lives of her heroines, her "gods." She sacrifices her actual bodily existence, including all sexual feeling and most of her other senses, because the waldo does not have enough bandwidth to allow her to experience usual sensations. She does all this so that she can be looked on with the admiration she thinks of the elite as having.

Tiptree's exploration of mimetic desire is not a minor part of this story; if it were, one would be more apt to blame technology. But the technology here is mainly an exceptionally powerful tool in the hands of the corporations that are trying to tap into this desire to imitate others. At several points, the narrator explains that GTX and other corporations that try to exploit these desires do not even try to understand how they work. They just know that they are powerful. The narrator tells her father to remember Jean Harlow. "A sexpot, sure. But why did bitter hausfraus in Gary and Memphis know that the vanilla-ice-cream goddess with the white hair and crazy eyebrows was *their baby girl?* And write loving letters to Jean warning her that their husbands weren't good enough for her? Why? The GTX analysts don't know either, but they know what to do with it when it happens" ("Plugged In," 62; emphasis in original).

The GTX analysts do not need to know, nor do they care, that the origin of the desire they fuel is self-loathing. Girard explains that the subject's real desire is that his own being, for which he feels revulsion, be completely absorbed into another. Tiptree literalizes this triangle as far as is possible by having P. Burke actually possess the body of the desired other through the use of the waldo. Although technology did not cause Burke's self-loathing, by offering the life of the waldo it feeds the illusion that one can draw ever closer to the possession of the lives of the beautiful people. In other words, it enables people to believe that this kind of transcendence, the virtual extension of one's own life into the life of another, will eventually become a fact, real enough to live by.

Another way to understand this point is to show how Tiptree anticipates Toni Morrison's story of Pecola Breedlove in *The Bluest Eye*. As I will discuss in chapter 4, Pecola, a little black girl, believes that gaining blue eyes would lead to her being loved. When Soaphead Church promises to give her the blue eyes, he validates Pecola's self-loathing. Pecola ends up destroyed by this validation; Morrison writes that "a little black girl yearns for the blue eyes of a little white girl, and the horror at the heart of her yearning is exceeded only by the evil of fulfillment."[32] Tiptree illustrates how technoscience validates self-loathing by making it literally possible for her protagonist to attain the thing that she thinks will cause her to be loved by others. She actually gives P. Burke the bluest eyes, and rather than solving her problems, it destroys her soul. As Girard explains, this mimetic desire is a "corrosive disease" that infects "the most intimate parts of being." And the worst thing about it, argues Girard, is that the self-alienation actually worsens as the distance between the subject and the object of her desire decreases.[33]

Both Morrison and Tiptree knew that it is tempting with stories like *The Bluest Eye* and "The Girl Who Was Plugged In" to blame the protagonists for their self-loathing, similar to the way readers usually blame Emma Bovary for hers. But between the nineteenth and twentieth centuries something significant changed. Emma chose to immerse herself in the worship of her heroines by entering particular novels, but in the media-saturated twenty and twenty-first centuries, "hero worship" more insidiously and invisibly affects everyone's view of everyone else. P. Burke is part of a culture incapable of seeing her misshapen body as beautiful. She is valuable only as a commodity, something of use to corporations that sell her image. The instrumentalization of the person that Habermas feared is now complete.[34]

The power of this narrative to challenge the prevailing paradigms is clear when we consider the story Tiptree could have told. Tiptree could have ended the tale with Paul discovering that the real P. Burke is beautiful, learning the classic lesson that "beauty is only skin deep." But she instead illustrates that the real P. Burke never had a chance because everyone, baited by the promise of technology, is breathing this air of mimetic desire, even those who seem to rebel against it.[35] Tiptree takes pains to tell the reader that Paul, the son of a GTX executive, fancies himself a rebel who is appalled at the world his father has made. But when he sees Delphi, he falls in love with her image, onto which he projects his recent reading of *Green Mansions*. When he sees Delphi "he sees Rima, lost Rima the enchanted bird girl, and his unwired human heart goes twang" ("Plugged In," 65). Paul may be unwired, but he is caught in the same trap of mimetic desire. P. Burke, five hundred feet under, thinks she is finally experiencing love—"the real hairy thing"— but what Paul loves is his vision of himself as the hero who saves the damsel in distress.[36] There is no chance that Paul would accept the real face of P. Burke because he did not even accept the real face of Delphi. Even if Burke could become Delphi, she would not be loved. The most tragic aspect of the story may be that Burke dies still believing otherwise.

Like many prophet-artists, Tiptree turns to extreme measures to get her readers to see the problem: the grotesque. Lesser writers might have been tempted to employ sentimentality, but this gritty story leaves no room for it. If Tiptree had wanted readers to feel sorry for P. Burke, she would have made her a more likeable character, more like the ugly duckling who is only temporarily ugly, as she is on the way to being "fixed." That is how the producers of *The Swan* clearly presented their ugly ducklings. Instead, the narrator seems to loathe P. Burke, calling her a "monument to pituitary dystrophy" that no surgeon would even bother to try to help. She has a half-purple jaw, a "jumbled" torso, and mismatched legs. She can barely walk without running into things. And Tiptree gives her no likeable personality traits; indeed, she doesn't really give her much of a personality at all. But Burke is a perfect example of the grotesque, an exaggeration meant to be a challenge to the readers' vision of the world by being simultaneously repulsive and attractive, like a train wreck. The narrator is giving this prophetic vision to a specific person who would deny involvement, as if to say, "You who are trading stocks, you think you are not complicit in this, but you are!" The reality is that in a world in which good looks are for sale, the technological "haves" will be increasingly separated from the "have nots," who become increasingly ugly in the comparison. The narrator chides her listener to look at the gods who are coming out of a store called Body East. "You don't believe gods, dad? Wait. Whatever turns you on, there's a god in the future for you, custom-made" ("Plugged In," 43). And this god will make people loathe their actual bodies into oblivion.

That consumer capitalism will always promote and never challenge the self-loathing of the vast majority of people is not an insignificant argument in today's bioethical debates. In his book *Radical Evolution,* Joel Garreau sets out three different scenarios vying for prominence in the biotechnological future.[37] The three scenarios are "Heaven," "Hell," and "Prevail." Garreau's "Heaven" scenario is best illustrated by the story of Ray Kurzweil, the famous writer of *The Age of Spiritual Machines.* Kurzweil is excited by the prospect of advances in biotechnology that will enable humans to take over their own evolution into higher states of being, into gods. "What we see in evolution is increasingly accelerating intelligence, beauty. . . . The evolutionary process . . . has continued with our technological growth of human cultural and technological history. . . . We see exponentially greater love."[38] This vision of the future is exactly what Tiptree challenges, because she sees that technological evolution is only going to be available to those who can afford it, at least initially, and that their desires will likely be much more base. Shows like *The Swan* make it all too clear: when people turn to biotechnology to improve themselves, they are trying to gain a competitive edge; they are not interested in learning how to love the other ugly ducklings. They want to learn how to leave the others behind, how to beat them in the competition. It is hard to imagine greater love evolving out of a system that thrives on its absence.

In this way, Tiptree's critique is deeper than that of Joss Whedon, whose television series *Dollhouse* ran for two seasons, from 2009 to

2010. Whedon probably borrowed ideas and terminology for his show from Tiptree: the dolls (called "actives") are controlled by "remotes."[39] But unlike Burke/Delphi, these dolls are actually beautiful young women and men who have had their own minds (temporarily) erased, subject to reprogramming at the whim of the current customer. The "volunteers" have been blackmailed, bribed, or otherwise manipulated into becoming dolls. Although Whedon does explore the ethical problems of using human bodies in a pay-for-fantasy scenario, the show's appeal derives largely from the audience's voyeuristic sharing of the fantasies of the customers, who get to do whatever they want with the beautiful body of Echo (Eliza Dushku) while her real mind is put on a shelf somewhere. Television viewers are, in fact, paying to experience these fantasies themselves. *Dollhouse* thereby explores but does not fundamentally challenge the pattern of mimetic desire. Even though the technology is just as far-fetched, the horror of "The Girl Who Was Plugged In" is much more real. The grotesque in this story functions, in Geoffrey Galt Harpham's description, as an equivalent to the "paradigm confusion" of metaphor, in which "unlike elements are yoked by violence together." The grotesque serves to cause a mental crisis "that we must suffer through on the way to the discovery of a radical new insight."[40] The story clearly expects to shock readers into seeing their complicity in the already-present horror of mass voluntary acquiescence, through consumer capitalism, to the illusory lives of the beautiful people.

When asked about her speculative novel *Oryx and Crake*, Margaret Atwood explained that novelists understand that the issue is not technology but desire. "The driving force in the world today is the human heart—that is, human emotions." But as *Oryx and Crake* suggests, as the tools become more powerful, the level of threat rises, and our responses become even more crucial and as easy as ever to exploit. "Our tools have become very powerful. Hate, not bombs, destroys cities. Desire, not bricks, rebuilds them. Do we as a species have the emotional maturity and the wisdom to use our powerful tools well? Hands up, all who think the answer is Yes. Thank you, sir. Would you like to buy a gold brick?"[41]

Although shows like *The Swan* are not going to lead to mass destruction, it is worth pondering how they influence our view of ourselves, our view of others, and our decisions on how to spend resources. As Martha Nussbaum reminds readers, these questions, which really boil down to the question, how should we live?, constitute the ethical nexus of philosophical and literary studies.[42] It is nothing less than the question of how to attain the good life. And though most people recognize the excesses of shows like *The Swan,* it takes a visionary like Tiptree to get us to ask, along with Pamela Orosan-Weine, "Does our acceptance of body alterations have any cultural or moral limits, or does it just slowly creep forward, advancing toward whatever technologies become possible? Without a religious mindset, how do we evaluate the question of whether these interventions are good for us?"[43]

The ironic thing about Orosan-Weine's assumption that religion does not or should not enter into our ethical framework is that there is, in fact, a "religious mindset" at play here. The set of *The Swan* is designed to look like everyone is gathered to worship the newly formed goddess, a goddess who is promised and then "revealed" at the end of the show.[44] The refreshing originality of "The Girl Who Was Plugged In" is its acknowledgment that image always involves illegitimate worship. And so Tiptree makes it impossible to miss that the goddess Delphi is an empty shell.[45] Like all celebrity worship, the object of the worship is an illusion. In a final irony, the narrator points out that the only person who was sad when the real P. Burke died was Joe, the technician who trained her, because "P. Burke, now a dead pile on a table, was the greatest cybersystem he has ever known, and he never forgets her" ("Plugged In," 78). Tiptree sacrificed P. Burke to insist that whatever the "real hairy thing" of genuine love for others may be, we need it in order to attain the good life. And like all good things, genuine love is shadowed by imposters and cannot be easily found.

CHAPTER 4

The Scorned People of the Earth
Reprogenetics and The Bluest Eye

> *The key difficulty in receiving the beauty of the world these days is*
> *that such teaching is rooted in the act of looking at the world as it*
> *is, while the dominant science is rooted in the desire to change it.*
> —George Grant, *Technology and Justice*

No one would deny that the biotechnological revolution has advanced with astonishing speed. DNA was discovered a mere fifty years before the human genome was completely mapped in 2003. In less than a century, corporate scientists have bioengineered a wide array of plants and animals: more nutritious golden rice, pigs with less toxic manure, cows with humanized milk. We inhabitants of the twenty-first century, as used to the daily development of new technologies as we are to the rising sun, seem barely to notice what is happening. In one generation, biology as a science has given way almost completely to biology as technology; as Gregory Stock describes it, "in one century, we have moved from observing to understanding to engineering."[1] The technology has opened as many ethical questions as it has new opportunities, so many that the average person is tempted to give up on answering them. Should we produce transgenic species of animals? Is it okay to develop the technology to grow human organs in pigs? Should athletes be allowed to enhance their performance with gene doping? Should we create headless human bodies in order to harvest organs?

Few questions raise more difficulties than those that touch directly on human reproduction. The new possibilities for parents to influence the genetic inheritance of their children have grown so quickly that Lee Silver, the prestigious professor of molecular biology and public affairs at Princeton University, has dubbed it the new science of "reprogenetics." With advancements in reproductive biology and genetics, Silver promises that we humans have now "tamed the fire of life" and gained the power to control our destiny as a species.[2] In reprogenetics, he adds, what had formerly only been the province of science fiction will become reality.

Pre-implantation genetic diagnosis, or PGD, is already relatively commonplace. PGD can be used by any prospective parents (with adequate means) who want to create embryos and then select from among them which ones to implant during the process of in vitro fertilization. Couples who carry the genes for Huntington's disease, for example, can create multiple embryos, screen them for the disorder, implant only the ones they choose, and discard the rest. But PGD is only the beginning of the dream of reprogenetics. The ultimate goal (still far in the future) is germline genetic engineering, whereby prospective parents can pick and choose from among a variety of genetic traits for their children. The goal, as Stock insists, is "self-directed evolution." In other words, germline genetic engineering will quickly move beyond therapy into being used primarily for enhancement.[3]

With the speed of these developments, one would think that caution and careful discussion would be everyone's default position. After all, with reprogenetics we are talking about changing someone else's life in a permanent way. Caution would seem to be especially warranted given the fact that we live in a culture in love with the glamour of technology and the promise of individual autonomy, at all times expressed in a consumer-driven marketplace. As Lauren Slater reported in 2002, for some people, the promise of technological enhancement is not just about improving intelligence or musical ability but about limitless possibility for human change. Joe Rosen, a plastic surgeon, told her that "the body is a conduit for the soul. . . . When you change what you look like, you change who you are."[4]

But caution has not proven to be Western culture's main response to any given technology, and this one will probably be no different. Proponents of genetic enhancement technology usually use its inevitable adoption as an argument in its favor. For instance, Ronald Green writes:

> I believe, with near certainty, that sooner or later we will begin to modify our genes and that we will survive doing so. . . . I believe that we are capable of bringing intelligence—"design" in the best sense of the word—to our reproductive lives. I am sure that we will make mistakes, most of which I hope we can correct and learn to avoid repeating. I also believe that eventually we will grow accustomed to a world where human beings make themselves physically and mentally better than they are today. Genetic science has opened our biology up to self-construction and directed evolution. We will certainly try to bring our biology under our control as we have done with so much of nature.[5]

Although Green is probably correct about the future of the technology, he knows that its inevitable use is not an adequate argument for moving ahead with genetic engineering. So he tries to answer objections to the technology that range from the problem of unintended consequences to the concern about the creation of a "genobility." Green, like other proponents of genetic engineering, reduces his opponents into two basic camps: those who, arguing from a theoretical or theological perspective, think it is inappropriate for humans to play God, and those who, arguing from a practical perspective, worry about some future implications. But it is not that simple. Neither of these camps is adequately concerned with what the desire for genetic engineering says about us as a society today or, more to the point, how our thinking about these technologies affects our current relationships and decisions. It is time that we changed the question away from parental rights and toward the question of attitudes and motivations. Specifically, it is time to ask, what happens to love, particularly parental love, in a culture that is so quick to grab at the prospect of genetic engineering? This is not a question that can be answered by a scientist or even an ethicist. To answer this question, we need narrative.

TONI MORRISON'S *THE BLUEST EYE*

While in elementary school, Toni Morrison met a black girl who said, with sorrow in her voice, that she wanted blue eyes. When Morrison pictured it, she was "violently repelled by what [she] imagined she would look like if she had her wish."[6] Years later, Morrison translated the questions she had about this girl's desire into her story about Pecola Breedlove, a little girl living in the 1940s who prays for blue eyes. Like all of Morrison's novels, *The Bluest Eye* is intently focused on the vision that one person has of another, insisting that the quality of our love is at the heart of all moral questions. As Morrison's narrator observes at the end of the novel, "love is never any better than the lover. Wicked people love wickedly, violent people love violently, weak people love weakly, stupid people love stupidly, but the love of a free man is never safe. There is no gift for the beloved. The lover alone possesses his gift of love. The loved one is shorn, neutralized, frozen in the glare of the lover's inward eye" (*TBE,* 206).

Toni Morrison's first novel is not specifically about technology or about the ethics of genetic trait selection. But it puts all the cherished assumptions of the reprogenetic agenda under intense pressure. What a reprogeneticist might call love—giving a child what you think she would want—it specifically names evil. Set in the context of a world that is instructing us to believe that selecting traits will give our children happiness, the novel is a potent corrective. It forces the ethical questions back on readers; it makes us question how and why we love the way we do. The novel does this by focusing our attention not on Pecola's choices but on the views of her held by those around her: parents, fellow students, and the entire community. By contrasting the love that the narrator, Claudia, learns to have for Pecola with the failure of Pecola's parents to love their own child (and of others to step in to help), it redirects our thinking. *The Bluest Eye* insists that learning to love others is always about becoming better people ourselves. It also dares to suggest that the more we long for new and improved children, the more poorly we may love the ones we actually have.

THE HORROR AT THE HEART OF HER YEARNING

The Bluest Eye intermingles narrative point of view between the first person and the third person: between a young neighborhood girl, Claudia MacTeer, and an unknown narrator, who is most likely Claudia years later. Claudia and her sister, Frieda, encounter Pecola as they are all trying to make sense of a world that calls Shirley Temple, but not black girls, beautiful. Pecola's family members, particularly her mother, have so internalized the white standard of beauty that they believe themselves to be ugly, and they heap that self-contempt upon Pecola. Pecola is raped by her father (Cholly), has a baby who quickly dies, and eventually goes to a neighborhood man named Soaphead Church who makes her believe he actually has the power to give her blue eyes. The resulting disjuncture between desire and reality becomes a kind of last straw for her, and she splits into two personalities, one trying to convince the other that she does have such beautiful blue eyes. Pecola never recovers and ends up a pariah in the community.

It is difficult to overstate how radical a story *The Bluest Eye* was and is. When Morrison began writing the novel in the late sixties, she did so in part in order to write the kind of story that she wanted to read but was not available. At a time when black girls are on the margins of society, Morrison puts two front and center: Pecola and Claudia. This is already a loving political and ethical act: it makes both of these girls—and all girls similar to them—visible to readers. In *The Human Condition,* Hannah Arendt explains that the telling of an individual's story is one of the primary ways that the individual can be "seen and heard," that he or she can make an appearance in public. Such a public appearance is central to experiencing reality as a human being. "To live an entirely private life means above all to be deprived of things essential to a truly human life. . . . The privation of privacy lies in the absence of others; as far as they are concerned, private man does not appear, and therefore it is as though he did not exist."[7]

Although Morrison does make Claudia and Pecola visible, like Ralph Ellison in *The Invisible Man,* the story that Morrison wants primarily to tell is about Pecola's invisibility. The novel famously begins with a quotation from the grammar primers common in the mid-twentieth century, the ones that tell the story of Dick and Jane: "Here is the house. It is green and white. It has a red door. It is very pretty. Here is the family. Mother, Father, Dick, and Jane live in the green-and-white house. They are very happy." The second time it is quoted, the punctuation is removed, and the third time, the words are jammed together "hereisthehouseitis greenandwhiteithasareddooritisverypretty . . ." and on to the end. As any student of sociology knows, a primer contains more than a grammar lesson; it subliminally teaches children what is normal. So Morrison's squeezing the words together is a kind of perversion of that message, an illustration of its choking notion of normalcy. Pecola is nowhere to be found in the world of Dick and Jane or any other cultural representations of little girls.[8] Indeed, Pecola is doubly invisible: first, to the white community, and second, to the black community, which ostracizes her family. When Pecola is brought to stay with the MacTeer family as a "case" child, Claudia explains that Cholly Breedlove's failures had put his family "outdoors," which was "the real terror of life." "Being a minority in both caste and class, we moved about anyway on the hem of life, struggling to consolidate our weaknesses and hang on, or to creep singly up into the major folds of the garment. Our peripheral existence, however, was something we had learned to deal with—probably because it was abstract. But the concreteness of being outdoors was another matter—like the difference between the concept of death and being, in fact, dead" (*TBE,* 17–18). Claudia knows that if the black community puts someone "outdoors," it is the final margin and as good as being dead. And Pecola is dead from the beginning, squeezed out of any real life.

Perhaps the most devastating critique the novel offers is the obvious fact that Pecola could have survived this communal ostracizing were it not for one thing: the failure of her own mother to love her. Pauline, Pecola's mother, has learned that she does not have any of the characteristics that the larger culture is calling beautiful, so she begins to live in a fantasy world of her own making, going to the movies more and more often. In a section devoted to Pauline's thoughts, she reflects that *"the onliest time I be happy seem like was when I was in the picture show. . . . White men taking such good care of they women, and they all dressed up in big clean houses with the bathtubs right in the same room with the toilet. Them pictures gave me a lot of pleasure, but it made coming home hard, and looking at Cholly hard"* (*TBE*, 123). The more she sustains the mental fantasy world, the more her real existence becomes a source of contempt. At the movies, she reflects: *"There I was, five months pregnant, trying to look like Jean Harlow, and a front tooth gone. Everything went then. Look like I just didn't care no more after that. I let my hair go back, plaited it up, and settled down to just being ugly"* (*TBE*, 123). Pauline's self-contempt substantially alters her vision of others; it practically guarantees that she would reject any baby that exhibits the offending traits. And sure enough, when the baby comes, Pauline reflects: *"Eyes all soft and wet. A cross between a puppy and a dying man. But I knowed she was ugly. Head full of pretty hair, but Lord she was ugly"* (*TBE*, 126).

Ignored by society, mocked by schoolmates, and living in a dysfunctional family that measures her by an impossible standard, Pecola wants to disappear. When she closes her eyes and concentrates, she finds that she can succeed in making various parts of herself disappear, everything but her own eyes, which "were always left." So then, she thinks, "What was the point? They were everything. Everything was there, in them. All of those pictures, all of those faces. . . . As long as she looked the way she did, as long as she was ugly, she would have to stay with these people. Somehow she belonged to them" (*TBE*, 45). So Pecola begins to pray for what she thinks is the only solution: to get blue eyes herself.

Pecola's desire for blue eyes is significant not just because her desire represents acquiescence to a white standard of beauty. Pecola had little choice in the matter; her mother, Pauline, had already succumbed to an "education in the movies," which taught her that she was not Jean Harlow or Claudette Colbert but that she needed to be in order to be worthy of love. Pecola's desire is significant because she believes that changing her appearance will make her family love her. As she fantasizes, she thinks that if "she looked different, beautiful, maybe Cholly would be different, and Mrs. Breedlove too. Maybe they'd say, 'Why, look at pretty-eyed Pecola. We mustn't do bad things in front of those pretty eyes'" (*TBE,* 46). Internalizing her family's self-contempt, Pecola longs for a quick and easy escape, the escape of a miracle. The narrator reflects that "thrown, in this way, into the binding conviction that only a miracle could relieve her, she would never know her beauty. She would see only what there was to see: the eyes of other people" (*TBE,* 47).

Pecola's desire for a miracle is what connects this novel to the question of the motives behind the use and development of enhancement technology. Pecola goes to the only person in town who she thinks can provide her with the needed miracle: Soaphead Church. The narrator remarks that Soaphead Church "was what one might call a very clean old man," a mulatto Anglophile who wants the entire black race to "marry up" into having lighter features. His Gnostic contempt for the body translates into pedophilia; not wanting the hairy and smelly bodies of real women, he preferred "those humans whose bodies were least offensive—children" (*TBE,* 166). It is in this character that careful readers can find a critique of the reprogenetic agenda, for Soaphead Church evinces how easily contempt for others can masquerade as love. He pities Pecola and wants so badly to give her what she wants that he promises the blue eyes, and then makes her believe she got them. "So it was. A little black girl yearns for the blue eyes of a little white girl, and the horror at the heart of her yearning is exceeded only by the evil of fulfillment" (*TBE,* 204). This one sentence is a brilliant analysis of the novel's contemporary relevance. The "horror at the heart of her yearning" has been explained by much commentary on this novel: the horrifying cultural reality that a little black girl is not loved for who she is, but would rather destroy herself in order to be accepted. But the evil of fulfillment, especially at the hands of a character like Soaphead Church, requires further discussion.

THE EVIL OF FULFILLMENT

I teach *The Bluest Eye* nearly every semester, and there is usually at least one student who gets Soaphead Church completely wrong. While most readers easily agree that a racist pedophile who claims he loves a little girl to whom he makes false promises (and whom he uses to kill a mangy dog) does not love her, these few students somehow read the novel and are able to come away from it thinking that, finally, we have a character who "really loves Pecola." The question of interest to me here is why: Why do these students believe him? And can we learn anything about the way we love others, especially all children, by studying this situation?

As I have mentioned, Morrison is intensely interested in the question of lovers and the quality of love each one has. She has said repeatedly that her stories are all about love and its absence. In an interview with Charles Ruas in 1981, she explained her particular interest in the question of love for children.

> Certainly since *Sula* I have thought that the children are in real danger. Nobody likes them, all children, but particularly black children. It seems stark to me, because it wasn't true when I was growing up. The relationships of the generations have always been paramount to me in all of my works, the older as well as the younger generation, and whether that is healthy and continuing. I feel that my generation has done the children a great disservice. I'm talking about the emotional support that is not available to them any more because adults are acting out their childhoods. They are interested in self-aggrandizement, being "right," and pleasures. Everywhere, everywhere, children are the scorned people of the earth.[9]

Whether one agrees with Morrison that children are the "scorned people of the earth," there is no doubt that children are a kind of cultural barometer for how well a society takes responsibility for its weakest members. Through our behavior toward children—and especially on behalf of them—we can discern our true motives and attitudes more readily.

It is at this point that proponents of reprogenetics might argue that the desire for genetic enhancement is, in fact, motivated by a great love for our children, a love that wants to give children a life with less suffering and greater potential for success. In his book *Babies by Design,* Ronald Green devotes an entire section to quoting studies that show how readily unattractive people are discriminated against, including one wherein a research team followed parents of children in grocery stores and found that parents of attractive two- to five-year-olds were much more likely to buckle their children into carts than were the parents of homely children.[10] He uses this example to explain that physical characteristics will be some of the first that parents choose for their children, all in the loving desire to spare them from this kind of treatment.

But this is precisely the point when motives must be analyzed the most carefully and the character of Soaphead Church is so revelatory. Through this character, Morrison illustrates one of her main concerns: that love has been hollowed out in Western culture. In an interview, she commented that "love, in the Western notion, is full of possession, distortion, and corruption. It's a slaughter without the blood."[11] What is even worse is that what we call love can sometimes be contempt masquerading in a cloak of tenderness. The reason why some of my students misread Soaphead Church is because he is tender. He wants to spare Pecola any suffering. The narrator explains that "as in the case of many misanthropes, his disdain for people led him into a profession designed to serve them" (*TBE,* 165). He had planned to be an Anglican priest but rejected that to be a social worker. Eventually that failed him, and he became a "Reader, Adviser, and Interpreter of Dreams," which is why Pecola seeks him out. And when she asks him to give her blue eyes, the narrator tells us that he "thought it was at once the most fantastic and the most logical petition he had ever received. Here was an ugly little girl asking for beauty. A surge of love and understanding swept through him, but was quickly replaced by anger. Anger that he was powerless to help her. Of all the wishes people had brought him— money, love, revenge—this seemed to him the most poignant and the one most deserving of fulfillment" (*TBE,* 174).

Soaphead wants to help her because *he agrees with her* that she is ugly. So love, in his definition of it, is not loving Pecola into redefining her sense of self, but taking steps that serve only to validate her self-contempt. His location of the problem is the issue. The problem, thinks Soaphead, is not that society and her parents have failed to love Pecola, but that she was born ugly and that no one has the power to fix her. So he pretends to take on that power for himself and then writes a letter to

God to inform him about all of God's failings. He tells God that he forgot about the children, that he let them "go wanting, sit on road shoulders, crying next to their dead mothers. I've seen them charred, lame, halt. You forgot, Lord. You forgot how and when to be God" (*TBE*, 181). This is not parallel to Morrison's complaint about the way children are being treated. It is instead the same kind of complaint that led Max More, founder of the Extropians (a group committed to reversing entropy through bioengineering), to read to an audience his letter to Mother Nature in 1992: "Mother Nature, truly we are grateful for what you have made us. No doubt you did the best you could. However, with all due respect, we must say that you have in many ways done a poor job with the human constitution. You have made us vulnerable to disease and damage. You compel us to age and die—just as we're beginning to attain wisdom. . . . What you have made is glorious, yet deeply flawed. . . . We have decided that it is time to amend the human constitution."[12]

Implicit in both of these letters is a contempt for the frailty of humanity and the belief that suffering has no value whatsoever. Soaphead thinks he would be better at being God because he understands that greater control over the messiness of life is both more beneficial and more beautiful. He believed that "something was awry in his life, and all lives, but put the problem where it belonged, at the foot of the Originator of Life. . . . Evil existed because God had created it. He, God, had made a sloven and unforgivable error in judgment: designing an imperfect universe" (*TBE*, 172). The narrator tells us that the "works he most admired were Dante's; those he despised most were Dostoevsky's" (*TBE*, 169). Soaphead wants things to be clean and tidy. He has use for a hierarchy in Dante's version of heaven and hell, but he has no use for spirituality that would lower itself to be expressed in the epileptic seizure of a character like Myshkin. Another way to say this is that Soaphead would approve of a God who renders judgments and makes things right, but not of one who would degrade himself to become a man through incarnation.

What kind of love Soaphead's view of the world strains itself into is made obvious even by his treatment of the mangy dog that he despises.

Since the dog was so revolting to him, he could not wait for him to die, but he "regarded this wish for the dog's death as humane, for he could not bear, he told himself, to see anything suffer. It did not occur to him that he was really concerned about his own suffering, since the dog had adjusted himself to frailty and old age" (*TBE,* 171).

In the same way, Soaphead thinks he is only concerned with Pecola's suffering. He tells God that "I did what You did not, could not, would not do: I looked at that ugly little black girl, and I loved her. I played You. . . . I, I have caused a miracle" (*TBE,* 182). Soaphead never acknowledges the convenience that this "miracle" was for him, that he could get Pecola to kill the mangy dog in the process. But what's worse, the miracle serves to be only the cause of Pecola's ultimate destruction. She creates a second self who sees her blue eyes and loves her. But of course, in this split between reality and fantasy, her true self is completely lost. As Morrison explains in the afterword, "she is not *seen* by herself until she hallucinates a self" (*TBE,* 215). It is after rendering the conversation between the two halves of Pecola's self that the narrator reflects. "So it was. A little black girl yearns for the blue eyes of a little white girl, and the horror at the heart of her yearning is exceeded only by the evil of fulfillment" (*TBE,* 204).

Although it is tempting to think that it is the false fulfillment of the promise that is the problem here, the "evil of fulfillment" of Pecola's yearning is much more elementary than that. By agreeing to give Pecola blue eyes, Soaphead Church validates and continues the core assumption that Pecola was ugly to begin with. If Morrison had written a science fiction novel in which Soaphead had the power to give Pecola the blue eyes, the evil outcome would be basically the same.[13] In fact, actually giving her the blue eyes might be worse, for it would confirm Pecola's conviction that she needs to change herself to be loved. And so it is with human enhancement, even when the changes are cosmetic; it contributes to an individual's dissatisfaction by misnaming the cause of the problem. Happiness—now fully dependent on others' views of us—becomes an infinitely receding horizon.

The fact that desire for something that cannot be truly attained yields only to infinite movement is the concern of Stephen Crane's poignant poem, "A Man Saw a Ball of Gold":

A man saw a ball of gold in the sky;
He climbed for it,
And eventually he achieved it —
It was clay.

Now this is the strange part:
When the man went to the earth
And looked again,
Lo, there was the ball of gold.
Now this is the strange part:
It was a ball of gold.
Aye, by the heavens, it was a ball of gold.[14]

Both the novel and the poem illustrate that the problem is not primarily with the technology itself, as if you could blame the creation of a faster ship for the speaker's desire to achieve the ball of gold, a ball that was delusional to begin with. But technology can contribute to the unhappiness behind a restless desire for change if it offers a continual succession of false fulfillments. In *The Bluest Eye,* Morrison is disgusted by the fact that the black community has sold its soul, in a way, to try to attain something that the white community also only desires and can never itself, by definition, attain. In reprogenetics, what parents think will make their children happy is misplaced at best and misdirecting at worst. If the genetic enhancement serves to give the parents the illusion that their children now finally have the means to happiness they never had, it contributes to a fantasy that will only, ironically, push their happiness even further out of reach. The knowledge that you were designed to have the talents of a concert pianist might make you that pianist, but what if they teach you to think, however accidentally, that fulfilling that destiny—and only fulfilling that destiny—will make you happy? It seems we have moved capriciously somewhere beyond the realm of parents saying to their children, "When I was your age, we used to have to walk to school—and it was in the snow. And it was uphill both ways." The difference between that kind of pressure to be grateful and the kind inherent in a possible parental remark like "I would have killed to have the genes I gave you" is a difference worth attending to.

I want to be clear that I am not making an argument about freedom or lack thereof in the children who would receive the genetic enhancements. I am talking instead about the perpetuation of an attitude toward one's body and toward happiness that ironically fuels dissatisfaction with ourselves and lovelessness toward others. This is Charles Rubin's main concern in his perceptive essay "Human Dignity and the Future of Man." Rubin explains how those who put their hope in human evolution through enhancement technology think that all that matters is that nobody would be forced to take on the enhancements—that one's free will would be respected. But to define the good life as that which can be gotten by enhancements or the "perpetual overcoming of the self in whatever manner the self wishes," is ironically most likely to result not in happiness but in a kind of enslavement, a "restless dissatisfaction or principled unhappiness." Even more important, argues Rubin, the definition of human dignity that such striving represents "characterizes no real persons or relationships, but rather is based on imaginative negation of the characteristics of real persons and relationships." This insight is precisely what *The Bluest Eye* substantiates, and it explains why we must allow the novel to shape our thinking about enhancement. Pecola's view of herself comes primarily from her parents and entails an utter negation of her very real body, its blackness, its funky uniqueness. While no one can prove that a genetically improved child will inherit such contempt for the flesh, it seems clear that people who choose to pay for enhancements will teach their children that technological solutions are the best way to solve all their "problems." The unenhanced child living next door might learn that same lesson with even more bitterness. The price of such a view, concludes Rubin, is that human dignity can "never flourish comfortably in any enduring here and now."[15]

A BETTER LOVER AND A BETTER LOVE

Reflecting later on her illustrious career as a novelist, Morrison was disappointed in *The Bluest Eye* because she felt that, in order to emphasize the destructive power of standards of beauty, she had to let those standards destroy Pecola. But of course, this does not negate the fact that Morrison's central move—to make visible the most invisible of all of society's members—was one of love. To render Pecola's life in the particular is to validate the right to existence and the essential beauty of all of the most vulnerable members of humanity. But Morrison also may not be giving herself enough credit for the way the story reveals the redemption of its part-time narrator, Claudia, and her sister, Frieda. Carl Malmgren makes a strong argument that although the narrative voice shifts between the young Claudia and a third-person narrator, it is Claudia who has put together the texts and Claudia's voice that gives the novel its unique power.[16] His argument is important because it is Claudia's act of storytelling that signals her acceptance of her moral agency.

As Lynne Tirrell explains, Claudia has a sense of self as distinct from others, a sense of self in time, and an ability to act intentionally, all of which are required for moral agency. But what seals the deal, in a sense, is learning to understand others, which requires the agent to "put their actions into the appropriate contexts and produce hypotheses about their reasons for acting. That is, one must give an account. A story is a special kind of account, for it recognizes and essentially uses the fact that the agent is a particular person living at a time within a particular society."[17] The reader can watch Claudia develop as a moral agent only through the narration itself; watching her grow to a different interpretation of the facts in her life is what gives *The Bluest Eye* its ethical power. In creating this tale, Trudier Harris explains, Claudia offers alternative possibilities to some horrendous circumstances, revealing the power of storytelling.[18] Because the novel is structured in a series of chapter flashbacks, it enacts and reflects the complex work of learning from the past that only narrative can provide.[19] The novel is thus a defense of the need for storytelling itself and argues for its indispensability in a world full of Soaphead Churches. As one critic puts it, the reader sees Claudia's mind grow, and "we may experience some growth of our own if we read carefully and well."[20]

One of the things that the reader learns is that Claudia and Frieda had been loved well (but not perfectly) by their mother and so had freedoms Pecola could never possess.[21] This love is a precise counter to the horror at the heart of Pecola's yearning, because it is love based in the here and now, in the real bodies of mother and daughters. In one of the most beautiful scenes in the book, Claudia tells the readers that when she was ill, she knew love even through the rough touch and angry words of her mother, who she says "despises [her] weakness for letting the sickness 'take holt.'" Claudia reflects that it "was a productive and fructifying pain. Love, thick and dark as Alaga syrup, eased up into that cracked window. I could smell it—taste it—sweet, musty, with an edge of wintergreen in its base—everywhere in that house. . . . And in the night, when my coughing was dry and tough, feet padded into the room, hands repinned the flannel, readjusted the quilt, and rested a moment on my forehead. So when I think of autumn, I think of somebody with hands who does not want me to die" (*TBE,* 12).

Claudia is a strong and insightful character who knows from the beginning that what she hates is not white people but whatever it was that made white dolls, but not black ones, beautiful and desirable in everyone's eyes. She never translates her dismay into self-hatred, in large part because she knows she is loved for who and what she is. What Claudia has to learn—and does learn—is how she and Frieda were complicit in Pecola's destruction. At the end of the novel, when Pecola is but a shell of a person, picking her way on the outskirts of town, the mature Claudia explains that they did not love her but that they had simply used her in order to see themselves as beautiful:

> The birdlike gestures are worn away to a mere picking and pluck-ing her way between the tire rims and the sunflowers, between Coke bottles and milkweed, among all the waste and beauty of the world—which is what she herself was. All of our waste which we dumped on her and which she absorbed. And all of our beauty, which was hers first and which she gave to us. All of us—all who knew her—felt so wholesome after we cleaned ourselves on her. We were so beautiful when we stood astride her ugliness. Her simplicity decorated us, her guilt sanctified us, her pain made us glow with health, her awkwardness made us think we had a sense of humor. (*TBE,* 205)

Claudia's moral imagination, shaped by her encounter with Pecola's suf-fering, has enabled a rare and wonderful self-knowledge. Claudia recognizes that her failure to love Pecola was born out of an essential selfishness that makes weak people dispensable, that views them as tools for the refinement of one's own self-image. She also recognizes that the resulting self-image is a false one, a set of fantasy virtues achieved only by way of comparison to the unvirtuous. She explains that "we were not strong, only aggressive; we were not free, merely licensed; we were not compassionate, we were polite; not good, but well behaved" (*TBE*, 205).

Claudia's moral imagination extends into territory far beyond this self-knowledge. So shaped have she and Frieda been by Pecola's story that they go against the tide of their entire community by loving someone even more unwanted than Pecola herself: Pecola's unborn baby. Everyone in the community wanted the baby dead, but these two girls self-sacrificially bury both the seeds they were trying to sell and the money they earned thereby in an effort to "make a miracle." Although their attempt at a trade with God is rooted in ignorance, Morrison uses this scene to relocate the whole idea of what is miraculous. Soaphead's miracle was to promise something that could only lead to destruction; Claudia and Frieda begin instead to believe in the power of love. The girls cannot make the baby live by burying money, but their love is, in fact, life-giving. In learning to love that which is not beautiful, they learn that love is what gives beauty. In a marvelous passage, Claudia is actually able to picture the baby in Pecola's womb: "It was in a dark, wet place, its head covered with great O's of wool, the black face holding, like nickels, two clean black eyes, the flared nose, kissing-thick lips, and the living, breathing silk of black skin. No synthetic yellow bangs suspended over marble-blue eyes, no pinched nose and bowline mouth. More strongly than my fondness for Pecola, I felt a need for someone to want the black baby to live—just to counteract the universal love of white baby dolls, Shirley Temples, and Maureen Peals" (*TBE*, 190). Although it was inevitable, in this novel, that such a baby would die (it completes Morrison's metaphor of the inhospitality of the environment), this passage demonstrates—and *also effects,* by way of its beautiful language—what Morrison wanted the novel to achieve: the redefinition of beauty that can only come from the beholder.

Claudia is a person of character, and it is translated into ethically responsible actions toward the other: the baby. This baby's very existence—especially the fact that she came unwanted—embodies all the "funk" that the "high yellow" girls and the Soaphead Churches of the world want to eliminate. With its woolly head, flared nose, and "kissing-thick lips," this is a defiantly black baby. What's more, Claudia emphasizes that it has a "living, breathing silk of black skin"—a beauty that derives from life-changing and growing as opposed to the artificially frozen, Pygmalion beauty of "synthetic yellow bangs suspended over marble-blue eyes."[22] At the beginning of the novel Claudia could only destroy the white dolls in anger. But here, Claudia's vision involves a complete move away from beauty as an ideal that pushes people more and more into artificiality, and toward an affirmation of life in all of its variety. This is beauty redefined and its beholder redeemed.

To put it in terms of the question of the ethics of human enhancement, the move Claudia makes in choosing to love the real baby's beauty over any ideal of beauty is to move her sympathies away from the ideal future baby and toward the actual present one. It mirrors Morrison's own reflection on the event that was the novel's genesis: listening to a young girl deny her own beauty. Shocked, Morrison reflects that "it was the first time I knew beautiful. Had imagined it for myself. Beauty was not simply something to behold; it was something one could *do*" (*TBE*, 209; emphasis in original). In philosophical terms, this is to move away from a Platonic ideal (with a fixed moral norm) and toward a more Aristotelian conception of the ethical as linked with teleology—that is, the process of a particular person aiming toward a good life. Paul Ricoeur's book *Oneself as Another* is indispensable in describing the import of the difference between these two views. Ricoeur turns to Aristotle's definition of the three types of friendship to begin to make his case—that friendship can be either for the sake of the good, or for the sake of utility, or for the sake of pleasure. Ricoeur uses this definition to explain how ethics must encompass the idea of mutuality, wherein each person loves the other for being the person that she is, and this is "precisely not the case in a friendship based on utility, where one loves the other for the sake of some expected advantage, and even less so in the case of friendship for pleasure."[23] Ethics requires mutual respect.

Using this type of friendship for the sake of the good as a base, Ricoeur develops what he calls the concept of solicitude. "Solicitude" means care for the other, but to exercise it requires that the self not view itself as greater than the other, as in, "I'm going to help you now, you poor sap." The dangers of this inequality in a caring relationship are nowhere more perfectly developed than in Gwendolyn Brooks's poem "The Lovers of the Poor," wherein the wealthy ladies from uptown want to give money to, but not actually mix with, the people they help. They judge who is worthy of their help, and the help is on their terms.

> Their guild is giving money to the poor.
> The worthy poor. The very very worthy
> And beautiful poor. Perhaps just not too swarthy?
> Perhaps just not too dirty nor too dim
> Nor—passionate. In truth, what they could wish
> Is—something less than derelict or dull.
> Not staunch enough to stab, though, gaze for gaze!
> God shield them sharply from the beggar-bold!
> The noxious needy ones whose battle's bald
> Nonetheless for being voiceless, hits one down.[24]

Unlike the behavior of these ladies, solicitude is not afraid of the return gaze of the other, because solicitude is a sympathetic exchange that establishes an equality of mutuality between two persons. Because solicitude requires sympathy, Ricoeur argues, it creates an equality that is established "only through the shared admission of fragility and, finally, of mortality."[25] In other words, solicitude, based on an exchange of giving and receiving, requires much more of the person than viewing the face of an other as that which demands justice, as is the case in the philosophy of Levinas. With this move, Ricoeur paints a picture of the self in relation to the other that reaches beyond the merely exterior demands of the face.[26] Solicitude requires a kind of humility, a recognition, in sympathy, that I could be that other person, that he or she is just as fragile and human as I am. Solicitude "adds the dimension of value, whereby each person is *irreplaceable* in our affection and our esteem."[27] This is what moves ethics beyond the mere performance of a dreary duty and into the realm of loving one's neighbor as oneself. In solicitude, ethics moves away from abstract concepts and toward taking responsibility for real people.

When these ideas are brought to the question of reprogenetics, their salience emerges. Everyone agrees that for many years to come, genetic engineering will proceed primarily by way of pre-implantation genetic diagnosis (PGD). Since parents will select some embryos based solely on the traits they possess, and discard the others, PGD is literally only possible when these parents have some notion of their children as replaceable. It is my contention that starting with this view impoverishes, even if seemingly only very slightly, our view of the other person as a gift, as someone who is irreplaceably important to us. The more we think of ourselves as the engineer of the other, the less we can identify with the undeniable fact that we were all once embryos ourselves. Sympathy decreases and, with it, the possibility of solicitude, which Ricoeur sees as the basis of ethics.

Ronald Green's attempt to answer this sort of objection is an example of how easily we can miss the subtle dangers behind the possibility of a whole culture changing its vision of the next generation. When Green describes how a team of researchers has recently discovered the genes that control skin color in Europeans and Africans, he concedes that the thought of selecting a child's skin color raises difficult questions. In response to Thomas Murray's questions, "Should we encourage biomedical fixes for complex social problems? Or would we be wiser to deal with the roots of prejudice?," Green argues that the questions are not new, but have been faced by Asian Americans who want to "Westernize" their eyes or by Jews who want to alter their noses. "In all cases," Green writes, "individuals and families had to ask whether it was better to seek relief from discrimination by fleeing one's differences or by staying and fighting for them. . . . Ultimately, the decision is highly personal, and one for which parents will probably come down on both sides."[28] This deflection to individual choice does not answer Murray's questions. Murray worries that to encourage biomedical fixes to the "problem" of race is to hack perniciously at the individual branches— not the communal roots—of the problem of prejudice. It is to turn away as a society from the difficult work of moral education and toward a seemingly easy solution that could easily backfire. If a community of people begins to put their energy behind giving children certain traits that the community perceives as desirable, is it not logical that children who do not exhibit those traits will become less and less desirable in that community's eyes? I agree with Green that genetic engineering will probably not diminish the loving care we give to our own, individual children, whether they be altered or unaltered. But I do wonder whether or not it will diminish the ability of future Claudias to love future Pecolas—and all of their unborn babies. Caution seems to be the only wise choice. If we do not become better at loving all children, genetic engineering might simply become the newest way that we ensure that children remain the scorned people of the earth.

PART
III

Posthuman

Language

CHAPTER 5

What Makes a Crake?

*The Reign of Technique and the Degradation of Language
in Margaret Atwood's* Oryx and Crake

> *The spare time of the* animal laborans *is never spent in anything
> but consumption, and the more time left to him, the greedier and
> more craving his appetites. That these appetites become more
> sophisticated, so that consumption is no longer restricted to the
> necessities but, on the contrary, mainly concentrates on the
> superfluities of life, does not change the character of this society,
> but harbors the grave danger that eventually no object of the world
> will be safe from consumption and annihilation through
> consumption.*
>
> —Hannah Arendt, *The Human Condition*

The first four chapters of this book have considered the ethics
of human enhancement technologies by considering individual choices
for improvement, whether it be by striving to become more beautiful,
more skilled, more authentic—or just happier. From "The Birth-mark"
to *The Bluest Eye,* each narrative revealed individual characters dealing
with their own dissatisfaction, desires, gropings, and failings.

For a large number of the writers I am calling "prophets of the post-human," individual choice is necessarily only part of the story. It is only the beginning. After all, some individuals will inevitably end up with more political power than Pecola Breedlove, Philadelphia Burke, and even Aylmer could ever hope to have. Furthermore, as I mentioned earlier, to have a prophetic voice in the biotechnological revolution is to attempt to reveal the "dominant consciousness" of advanced technological societies and to offer and nourish alternatives. But since the creation of the atomic bomb, that prophetic voice has been somewhat circumscribed by the overriding and relatively new fear that humanity's ultimate destiny is literally at someone's fingertips. It is thus the fear of totalitarian rule that has driven the creation of some of the most compelling dystopic novels of the twenty and twenty-first centuries, including Huxley's *Brave New World,* Atwood's *The Handmaid's Tale,* and even the recent young adult best seller Suzanne Collins's *The Hunger Games.*[1]

As interesting as these novels are, their ills are mostly caused by nefarious governments, a fact that limits their usefulness when it comes to recent debates over the ethics of enhancement technology. Even in *Brave New World,* which most scholars argue represents a more plausible dystopia than many, the socioeconomic situation is the result of totalitarian control.[2] Totalitarian social engineering of any sort is certainly a worthy opponent, and few can argue against the desire to keep power switches away from Dr. Strangelove. But critical appropriation of these narratives has permitted proponents for unfettered development of bio-enhancement technologies to set up a straw man. Ronald Bailey, for example, dismisses bioethical warnings derived from *Brave New World* by arguing that since Huxley is worried about totalitarianism and not the free use of technology, the way to avoid his nightmare is to promote and protect liberal democratic practices. In fact, argues Bailey, the best way to do that is to give the biotechnological revolution full sway. He argues that "the future toward which the biotech revolution is taking humanity is, in fact, almost the exact opposite of the Brave New World," because what the "modern technologies all have in common is that they give *individuals* more choices and options over how to enhance their lives, improve their health, and bear children."[3] If individuals have more options, the thinking goes, then democracy must be thriving, and there is nothing to worry about.

An opposite, but related, straw-man approach to these dystopic narratives is to ascribe to the artists a neo-Luddite fear of technology, as if they all believe that technology has a life and will of its own, without any human agency behind it. This is how Ronald Green, the author of *Babies by Design,* blithely dismisses Margaret Atwood's novel *Oryx and Crake.* Green, a bioethicist, believes that we should be free to make deliberate interventions into our genetic makeup for both therapeutic and enhancement purposes. "For most of our history, we have been the passive subjects of change," he writes. "In this new era we will take the direction of our evolution into our own hands." Green thinks that artists like Atwood are simply part of an unthinking "coalition of opposition" that has slowed down our progress, because she employs scare tactics in a novel in which "bioengineering has produced an apocalypse." Like so many other writers who apparently simply do not want anyone to tamper with nature, Atwood "criticizes our excessive love of science and our environmental intrusiveness. Her fear is that we can never retain full control of our creations."[4] Green thus lumps Atwood's critique into a pile with other sci-fi romps like the Terminator or Frankenstein movies, in which all the action begins after the decision to employ the technology has already been made, and the technology takes over.

Margaret Atwood's *Oryx and Crake* cannot be dismissed by either of these approaches. While it is true that *Oryx and Crake* describes an apocalypse, what Atwood fears is not total destruction brought about by technology run amuck or by a totalitarian government led by extreme outliers. Like all good speculative fiction, *Oryx and Crake* digs deeper. Its mode is primarily satirical; in this novel, the apocalypse is devised more to reveal society's current choices than to predict its inevitable future. The novel insists that it is not bioengineering that could cause our self-destruction but the continuation of a culture that encourages people to think about the purpose of human life in a narrow and nefarious way. Crake is Atwood's mad scientist in this novel, and although his character is exaggerated, it is recognizable. His views, though radical, are chillingly logical in a late modern technocracy that increasingly tries to *make* solutions to its problems. For Atwood, reliance on technique and process has been concomitant with a disintegration of language that can be seen in the degradation of the arts. This cultural change forms a society in which someone like Crake, a narcissistic technocrat with no regard for others, no capacity for love, can be elevated into a position of influence. With their destructive potential in full view, Atwood's novel encourages readers to return to the source and ask: What makes a Crake?

CRAKE THE MAKER

Oryx and Crake tells the story of two friends, Jimmy (Snowman) and Glenn (Crake), who grow up in the novel's immediate past—North America's near future—around the year 2025.[5] The novel's present world takes place after an apocalyptic destruction effected by Crake, a scientist bent on the fabrication of a new species of hominids called Crakers. Crake unleashes a powerful virus that destroys most of the world's population, so that all that is left is the Crakers, a handful of human survivors, and a large, genetically altered animal population that has become violent and predatory. Jimmy (Snowman) survives the blight, and we follow him as he tries to navigate the postapocalyptic world. Into Jimmy's survival story, by a series of flashbacks, Atwood interweaves the tale of how he got there: what he and Crake did as young boys, where they went to school, how Crake developed, and so on. It is one of Atwood's particular specialties, developed over many previous novels, to juxtapose past and present events powerfully. In this case, the technique works to give us two futures: the one we already inhabit and are moving closer into, and the one that could be our reality if nothing changes. It is the near future that is of the greatest interest to her and that provides a chilling warning: the near future of this novel is not so far from the North America of the present.

As readers slowly get to know the world in which Jimmy and Crake grew up, we learn that the sociopolitical landscape has been divided between the "Compounds" and the "pleeblands." The Compounds are highly organized and environmentally protected spaces where the real political power resides in corporations like HelthWyzer. The pleeblands, in which most of the population resides, are urbanized, more free, environmentally degraded, and less secure. Anything meaningful in the culture comes from the Compounds, where an ethic of efficiency and profit margin clearly reigns. The economic culture of the Compounds, which exploits the desires of people there and in the pleeb-lands, rewards technological innovation that boosts the bottom line. The most powerful companies are specifically exploitative: ReJoovene-scene develops fountain-of-youth products like AnooYou, and OrganInc Farms is responsible for *sus multiorganifer,* informally known as "pig-oons": large pigs in which human organs can be grown. The corporate culture clearly favors "numbers" people. Only scientific and techno-logical innovators like Crake may attend highly competitive and well-funded schools. Those without this kind of intelligence, like Jimmy, must attend the increasingly underfunded schools for the arts. Either way, only the most privileged have access to higher education; the rest are consigned to a life of meaningless consumption in the pleeblands.

Atwood's speculation of an ultimate division between the corporate haves and the plebian have-nots is not unusual. But Atwood's particular take on how such a thing could happen—and indeed is happening already—is uniquely illuminating. Her scenario extends the reasoning begun by Hannah Arendt in *The Human Condition.* Arendt tells the story of Western culture as one in which individuals' capacity to act po-litically in the public sphere has been greatly diminished. This dimin-ishment has been achieved, in part, through the ascendancy of *animal laborans,* or a definition of the human as a laborer who merely contrib-utes to the constant push against necessity, rather than as a doer who acts meaningfully in the public sphere.[6] Arendt's description of this as-cendancy is complex but can be described by the following narrative: When the market economy expanded, everything became an object of production and consumption, and when that happened, the values of productivity and abundance superseded other values such as freedom and plurality. The private world, dominated by values associated with production and consumption, in effect bled into and then obliterated any meaningful public space. How much of Arendt's story is accurate is certainly up for debate, but Atwood's Compound/pleebland vision seems quite consonant with it.[7] Government has collapsed into a de-centralized rule by corporation, the directives of which are handled by CorpSeCorps, a kind of police organization. The corporations have succeeded in reducing people to cogs in the machine. Both the pleeb-landers and the Compound dwellers have become mindless consumers, following only their own appetites. The things the corporations make are dictated by efficiency—ChikiNobs, the result of chickens that are genetically altered to be all breast and no head—or by the desire to regenerate oneself in the action of living to consume—the products of AnooYou promise NewSkins to replace old. Children exist to fulfill the desires of the parents: you can order an infant to specification from Perfectababe, Infantade, or Foetility. In all, this near future is clearly a "ferocious satire on late modern American capitalist society."[8]

The comparison gets particularly interesting when Arendt outlines the inevitable results of a society that thinks of itself primarily in terms of producing and consuming things. Everything, even the structure of society itself, becomes something that humankind makes, so that to be modern means to think in terms of *making* history, politics, government, and even, ultimately, our own happiness. Although early utopian political thinking contained the seed of this transition, its ultimate fruit came later, with the glorification of violent revolution as the only way to "make" a new body politic.[9] For Arendt, this kind of thinking opened up the "principle of utility" that enabled the development of utilitarian ethics, which is morality reduced to the attempt to quantify pleasure and pain. It is not surprising to find technology revered in cultures in which the principle of utility reigns; individuals now think of themselves as makers "of tools to make tools" that will lessen pain and effort and increase happiness.[10]

When a culture interprets acting in terms of fabricating, the move is not without great cost. Modern science and technology, argues Arendt, act directly into the processes of nature, so that what is done (made) can only be undone (unmade) by destruction. Other, more salutary ways of acting have been lost. "Nothing appears more manifest in these attempts than the greatness of human power, whose source lies in the capacity to act," writes Arendt. Because of the potential permanence of the action, the attempt "inevitably begins to overpower and destroy not man himself but the conditions under which life was given to him."[11] This argument sounds like the outline for *Oryx and Crake*. The problem is not that there is a behind-the-scenes dictator like Mustafa Mond in *Brave New World* but that widespread acquiescence to a definition of life as that which we make has enabled the rise of unheralded corporate power. Corporations, not government, are the makers and the actors. CorpSeCorps acts often and decisively to protect the corporate structure: they killed Glenn's (Crake's) father when he tried to act outside of their mandates; they are responsible for Jimmy's mother's disappearance as well. The world that Crake grows up in and is formed by encourages him to think of change only in terms of buying and selling, and of destroying and making anew. Crake and his vision for a new species he thinks can be easily made happy and free from desire are thus not surprising. Crake the Maker simply genetically fabricates beings so purely instrumental that Jimmy would later describe them as "animated statues."[12]

THE DEGRADATION OF LANGUAGE

Atwood's speculation about the near future of a society defined by making has even more resonance when she turns her observations to the ultimate fate of language and the arts in such a society. Jimmy and Crake grow up in a world that clearly marginalizes language and the arts so that they are merely tools of production and consumption, advertising or entertainment. Much of Atwood's humor is at work here: Martha Graham, the school in which Jimmy enrolled only because he wasn't smart enough for Watson-Crick, is underfunded and falling apart.[13] One can get a degree in Pictorial and Plastic Arts, and the curriculum consists of classes such as Image Presentation and Webgame Dynamics (*OC*, 188). Anything else in the curriculum is a joke, a remnant of a previous time. As Jimmy considers it, "a lot of what went on at Martha Graham was like studying Latin, or book-binding: pleasant to contemplate in its way, but no longer central to anything, though every once in a while the college president would subject them to some yawner about the vital arts and their irresistible reserved seat in the big red-velvet amphitheatre of the beating human heart" (*OC*, 187). Everything has been changed to reflect that language must be useful or it is thrown out: Martha Graham's motto had once been "*Ars Longa Vita Brevis*"—art is long, life is short—but is now the vapid "Our Students Graduate with Employable Skills" (*OC*, 188).

Jimmy and Crake's world represents what happens to art when society attempts to reverse the original motto of the Graham school to "art is short, life is long." The only artist figure in the novel is Amanda Payne, one of Jimmy's girlfriends. Her art consists in spelling out a single word on the grass using animal carcasses—an inevitably short-lived display. When the vultures come, she photographs the word from a helicopter. She chooses only words with four letters—"pain," "whom," and "guts"—and literally enacts their animation and execution. "It was a powerful process—'Like watching God thinking,' she'd said on a Net Q&A" (*OC,* 245). The arts are consigned to a negative reenactment of the corporate "vulturization" of language and, with it, of the soul itself. It is no surprise that Crake tries to eliminate the art impulse from his genetic creation of the Crakers, for he had fully internalized these values, claiming that art is only "an empty drainpipe. An amplifier. A stab at getting laid" (*OC,* 168).

Besides Amanda Payne, Jimmy is as close to an artist figure as can be found in *Oryx and Crake,* and much of the novel's moral energy comes from the stark contrast between him and Crake. But while Crake is clearly designed to stand for the utilitarian scientist type, Jimmy is not some simple savior of the human language, the English major's hero. Though Jimmy is inclined more naturally toward language, he, too, has been shaped by a culture that does nothing to encourage it. So Jimmy learns to manipulate language, to use it to gain power, money, or friends. When Jimmy graduates, he gets a job in advertising at AnooYou. He is told to "describe and extol" what people could become by using the company's products. "Hope and fear, desire and revulsion, these were his stocks-in-trade, on these he rang his changes. Once in a while he'd make up a word—*tensicity, fibracionous, pheromonimal*—but he never once got caught out. His proprietors liked those kinds of words in the small print on packages because they sounded scientific and had a convincing effect" (*OC,* 248–49).

The fact that Jimmy can easily make up words without being noticed means that nothing more is expected of language beyond serving as slogans. This phenomenon has been clearly described by the prominent mid-century sociologist Jacques Ellul:

In slogans, words are completely stripped of their reasonable and meaningful content. All oral propaganda rests on the fact that language loses its meaning and retains only the power of inciting and triggering. The word has become mere sound: pure nervous excitation, to which people respond by reflex, or because of group pressure. If a speaker fails to make use of the magic words which automatically stir up hatreds, passions, mobs, devotion, and curses, the rest of his language dissolves, as far as his listeners are concerned, into a gush of lava, an overflow of monotony, a contemptible fog that prevents or smothers action. The word thus loses its power.[14]

Words employed merely for their effect end up without any power at all. Jimmy knows this by instinct, but his cultural conditioning has been so thorough that he has no recourse. When he tries to tell a woman that he loves her, that she's the only one, he reflects that "she isn't the first woman he's ever said that to. He shouldn't have used it up so much earlier in his life, he shouldn't have treated it like a tool, a wedge, a key to open women. By the time he got around to meaning it, the words had sounded fraudulent to him and he'd been ashamed to pronounce them" (*OC,* 114). Jimmy knows that something has gone very wrong here.

Because Jimmy lives in a world of devalued words, Atwood's novel is itself an effort to demonstrate what is sacrificed thereby. Atwood's language glimmers in contrast to that of this flattened, colorless world. Jimmy's reflections on his dissatisfaction with language become, ironically, an opportunity for her to show the real power of language through the concrete edges of metaphor: "He knew he was faltering, trying to keep his footing. Everything in his life was temporary, ungrounded. Language itself had lost its solidity; it had become thin, contingent, slippery, a viscid film on which he was sliding around like an eyeball on a plate. An eyeball that could still see, however. That was the trouble" (*OC,* 260). Atwood's language is deliberately solid and visceral. It is grounded and powerful. When she describes how the older Jimmy has lost even the odd comfort he had found in words as a child, she uses an edgy metaphor: "It no longer delighted Jimmy to possess these small collections of letters that other people had forgotten about. It was like having his own baby teeth in a box" (*OC,* 261).

When the apocalypse occurs and Jimmy becomes Snowman, it is no surprise that he can still only grasp at words as sounds that have now become even more worthless. He knows he should hold onto them but cannot understand why. He is plagued by language, but almost none of it is his own. Most of what he hears in his head is completely appropriated discourse, dialogism gone amuck. In virtually every scenario in which he finds himself, the contrast between the appropriated, second-hand discourse and the raw needs of the moment is jarring. In one scene he tries to explain to the Crakers why he was talking to himself. He tells them he was talking to Crake. "What are you telling to him?" they ask.

> What, indeed? thinks Snowman. *When dealing with indigenous peoples,* says the book in his head—a more modern book this time, late twentieth century, the voice a confident female's—*you must attempt to respect their traditions and confine your explanations to simple concepts that can be understood within the contexts of their belief systems.* Some earnest aid worker in a khaki jungle outfit, with netting under the arms and a hundred pockets. Condescending self-righteous cow, thinks she's got all the answers. He'd known girls like that at college. If she were here she'd need a whole new take on *indigenous.* (*OC,* 97)

The dialogue in Snowman's head illustrates exactly how language has been increasingly devalued through the "official style," or what one might read in an academic textbook. Only people who have been severed from real contact with others (people who think that a "khaki jungle outfit" makes one a capable helper) use phrases like "indigenous peoples" or "contexts of their belief systems." Wendell Berry dubbed this kind of speech "tyrannese."[15] The postapocalyptic world merely punctuates the fact that such language is impoverished to the point of worthlessness. And even now, Snowman cannot completely escape it, though the beginnings of a regeneration of a more concrete and honest language ("self-righteous cow") do peek out.

Oryx and Crake is thus not primarily a novel about the dangers of specific technologies. It is a satirical exaggeration of what happens when a society, in submitting to the reign of technique, allows its language to become impoverished and its arts obliterated. This particular result had been noted well before Atwood satirized it. In *The Technological Society*, Ellul defined an advanced technological society as one that submits all aspects of life to the goal of greater efficiency.[16] In his later book *The Humiliation of the Word*, he explains how this goal affects language. The desire to be efficient dictates the need to grasp things quickly, to use language as a tool in the service of goals. Such speed, of course, works against the arts, which are profoundly inefficient and not easily "used." What replaces the richness of the arts are surface images and slogans, which are easy to grasp. As a result, slower, word-based thinking gets edged out of the way: "Images are the chosen form of expression in our civilization—images, not words. For though our era speaks, and abounds in printed paper, so that written thought has never been as widespread as today, still there is a strange movement that deprives the word of its importance. Talk and newspapers are like word mills to which no one attaches any importance anymore."[17]

Ellul is not arguing that words are inherently more important than images. He is referring to the kind of thinking that emerges when language is treated like a surface image. Users of impoverished language come to view the truth the same way: as reduced to that which can be easily verified, that which appears to be true. Because this thinking is based on an image that "gives rise to a feeling of evidence and a conviction that it is not based on reason," it seems to be unquestionable. That which "produces immediate assent cannot bear the discussion process. The conviction acquired in this way can only be attacked on its own ground: by other images and other 'evidences.'"[18] What's worse, explains Ellul, is that this kind of thought is always "committed" in that it encourages immediate action and reaction to the image instead of a careful discussion of options.

Ellul is certainly not the only one to argue that when language is impoverished, discourse degenerates into a kind of zero-sum game. It was also the thesis of Toni Morrison's 1993 Nobel prize acceptance speech. Dead language, Morrison proclaimed, "actively thwarts the intellect, stalls conscience, suppresses human potential. Unreceptive to interrogation, it cannot form or tolerate new ideas, shape other thoughts, tell another story, fill baffling silences."[19] In short, the degradation of language is an open door for violence against others.

TECHNONARCISSISM

For Atwood it is language and the arts that constitute our humanity. In a speech about the novel, Atwood explained that "'The arts'—as we've come to term them—are not a frill. They are the heart of the matter, because they are about our hearts, and our technological inventiveness is generated by our emotions, not by our minds. A society without the arts would have broken its mirror and cut out its heart. It would no longer be what we now recognize as human."[20] *Oryx and Crake* illustrates that true empathy—that which leads individuals to take responsibility for one another—is only possible in a society that protects language and the arts. The consequences of ignoring this fact has been a central concern of Atwood's since *The Handmaid's Tale*. Whereas that novel was a dystopia clearly in the tradition of Orwell's *1984,* in which the government literally controlled language, *Oryx and Crake* is a satire depicting how a culture's freely made choices can lead to what amounts to the same thing: empathy erodes; violence increases.

Jimmy and Crake grow up in a world in which the government need not prevent meaningful discourse because the internet has already done it for them. Their high school education in the Compounds has become a joke, and they learn what they learn from constantly surfing the net. When they surf, they are driven purely by immediate response to images and sound bites—they consume, use up, and quickly move on. Atwood's satire is at its best when exaggerating the empathy-degrading potential of the internet. The boys see Noodie News, which "was good for a few minutes because the people on it tried to pretend there was nothing unusual going on and studiously avoided looking at one another's jujubes" (*OC,* 81–82), and Felicia's Frog Squash, which was fun but quickly grew to be tedious because "one stomped frog, one cat being torn apart by hand, was much like another." With information and words everywhere, the boys have long given up on any arbitration between the important and the unimportant, the real and the virtual, the human and the animal. The internet here is for entertainment: what the sites share is an appeal to the instant gratification of libertine consumption. As Ellul had put it, "the spectacle-oriented society makes a spectacle of itself, transforming all into spectacle and paralyzing everything by this means."[21] The boys become voyeurs on the outside of everything, interacting with people who expect nothing from them beyond their momentary attention. Pornography is, of course, ubiquitous, but in Atwood's satire it is not the worst on offer. Instead, the internet abounds with socially acceptable public spectacles of violence. There is hedsoff.com, shortcircuit.com, brainfrizz.com, and deathrowlive.com, all of which show live executions. Viewers are tantalized by the promise that they are seeing a real execution but are also permitted to dismiss any uncomfortable emotion by retreating into the belief that it is all made up. "Crake said these bloodfests were probably taking place on a back lot somewhere in California, with a bunch of extras rounded up off the streets" (*OC,* 82).

With the line between virtual and real erased, no one need take ethical responsibility for actions performed against people.[22] The human body becomes for Jimmy and Crake merely an object for passive consumption. But Atwood's scenario does more than argue that the content of the internet commodifies people. In changing the way the content is accessed, the internet teaches a certain way of thinking. As Nicholas Carr explains in *The Shallows,* the internet provides no incentives to think in terms of wholes, backgrounds, or environments. "We don't see the forest when we search the Web. We don't even see the trees. We see twigs and leaves."[23] The way that Jimmy and Crake use the internet shapes the way they see people: as interchangeable parts. "So they'd roll a few joints and smoke them while watching the executions and the porn—the body parts moving around on the screen in slow motion, an underwater ballet of flesh and blood under stress, hard and soft joining and separating, groans and screams, close-ups of clenched eyes and clenched teeth, spurts of this or that" (*OC,* 86). The boys switch compulsively back and forth between spectacles, so that "if you switched back and forth fast, it all came to look like the same event. Sometimes they'd have both things on at once, each on a different screen" (*OC,* 86). The media itself mitigates against the boys seeing other people contextually, as a part of larger narrative. They are, instead, to be broken apart and consumed.

While Crake seemed to be unaffected by these sessions, Jimmy would "wobble homewards" sick from the dope and feeling "as if he'd been to an orgy, one at which he'd had no control at all over what had happened to him. What had been done to him. He also felt very light, as if he were made of air; thin, dizzying air, at the top of some garbage-strewn Mount Everest" (*OC*, 86–87). Jimmy feels his own moral weightlessness and is unwittingly bothered by it; he is never settled with this easy view of others as mere products for consumption. Atwood seems to suggest that the reason Jimmy's response differs from Crake's is because he is still responsive to language that challenges him. Jimmy's first encounter with Shakespeare is a case in point. Jimmy and Crake often frequented a site called "At Home With Anna K", in which a "self-styled installation artist with big boobs . . . wired up her apartment so that every moment of her life was sent out live to millions of voyeurs" (*OC*, 84). Jimmy first hears the language of Shakespeare not in high school, but because Anna K would sometimes read parts of old plays out loud while sitting on the toilet. During her rendition of *Macbeth*, Crake immediately wants to change the channel, but Jimmy insists that he wants to keep watching, though he does not even know why. He had "been seized by—what? Something he wanted to hear" (*OC*, 85). Jimmy does not even have adequate language to understand what he is missing, but he knows that this play, a play about the moral consequences of Crake-like ambition, is somehow important.[24] In this world, Jimmy's response to language, as minimal as it is, is the best available. So Jimmy becomes the novel's ersatz moral sun, emitting light like a star that has already burnt out but lives on in borrowed time.

The moral energy of the novel boils down to Jimmy's first encounter with Oryx. The scene stands out to Jimmy and to readers because of the rare rawness and three-dimensionality of its emotion. Jimmy and Crake are surfing as usual, visiting the porn sites Superswallowers, Tart of the Day, and finally HottTotts, a global "sex-trotting site" located in countries where children were cheaply purchased and filmed performing obscenities with disguised Westerners. Jimmy and Crake had seen countless naked little girls through this site, girls who remain unnamed and uninteresting to them except to fulfill a passing voyeuristic whim. But then Jimmy sees the little girl he would come to believe was Oryx: "None of those little girls had ever seemed real to Jimmy—they'd always struck him as digital clones—but for some reason Oryx was three-dimensional from the start" (*OC*, 90). She and a group of other little girls were licking a man covered with whipped cream when Oryx paused, smiled, and "looked over her shoulder and right into the eyes of the viewer—right into Jimmy's eyes, into the secret person inside him. *I see you,* that look said. *I see you watching. I know you. I know what you want*" (*OC*, 91). For the first time, the tables are turned on Jimmy, and he is the object of someone else's vision. As a result, two people are newly reified in the scene: Oryx, who is now a person, something more than an image, and Jimmy, who is no longer permitted to be merely a transparent eyeball. Jimmy is awakened to this sensibility by the appearance of a real face.

Jimmy's reaction to Oryx's face evokes Emmanuel Levinas's discussion of ethical responsibility to the Other.[25] Jimmy and Crake, by becoming voyeurs, were attempting to absolve themselves of any responsibility for the injustice represented by third-world cultures that would sell girls like Oryx into the porn industry.[26] For this voyeurism to succeed, the victims must remain silent and impersonal. But, like it or not, when the face appears, it makes a claim, specifically because *it speaks*. Language presupposes a plurality; in a way it makes a person leave her own home and engage with others. This engagement necessarily comes at some personal cost. As Levinas explains, "to speak is to make the world common, to create commonplaces. . . . It abolishes the inalienable property of enjoyment. The world in discourse is no longer what it is in separation, in the being at home with oneself where everything is given to me."[27] In the world of *Oryx and Crake,* language—in this case, the look that talks—is that which challenges the autoeroticism of the voyeur.

Put simply, by imagining that he hears Oryx speaking, Jimmy enters into conversation, and conversation implies plurality. Ethics has no meaning without true plurality. This is why for the moral weight of this scene to be complete, Jimmy must eventually find the real person who is this girl, and he does (or believes he does): Oryx. When he finds her, he listens to her whole story, and though he can never be sure what of her tale is factual, she certainly becomes more whole with each revelation.[28] Through his conversations with Oryx, Jimmy eventually comes to take a modicum of personal responsibility for joining in with the voyeuristic culture, but the lesson is not easy for him. In one of their most dramatic conversations, Jimmy, who has been in love with Oryx from that first encounter, is trying to convince himself that Oryx's previous life as a child porn actor never actually involved any sex. He really wants to know what she did, but she is never forthcoming and always tries to change the subject: "Why did this drive him so crazy? 'It wasn't real sex, was it?' he asked. 'In the movies. It was only acting. Wasn't it?'" And Oryx replies, in a sentence that is as close to Atwood's own voice as it gets, "'But Jimmy, you should know. All sex is real'" (*OC,* 144). The house of cards that had been Jimmy's moral indifference to internet exploitation of others begins to teeter.

Jimmy's moral growth is slight, and one wonders if it will be enough to mean anything in the postapocalyptic world he must learn to survive. But the tough vision of *Oryx and Crake* is that technocratic values produce educational environments in which a sensitive child like Jimmy is the exception, not the rule. Crake, the boy who would grow up to destroy nearly all of humanity, is the most likely product of his environment. Crake has been so morally deadened that neither internal nor external forces can compel him to take responsibility for particular other people. The face of Oryx means nothing to him, even after he has located her (or someone who looks like her) and has her brought to him. Atwood emphasizes the difference between Jimmy's encounter with her face and Crake's: in the above scene, in which Crake takes a snapshot of the website and files it indifferently away, but even more so when Jimmy later finds that Crake has been using that image as computer wallpaper, a portal to enter into his game playing of Extinctathon. Jimmy, who has never told Crake that the girl's image had moved him, feels violated: "It was a private thing, this picture. His own private thing: his own guilt, his own shame, his own desire" (*OC,* 215). Crake uses the image as a literal gateway to the playroom, when the image had been a much more important gateway for Jimmy, a gateway to greater responsibility for other people. In Ellul's terms, Crake's gateway is one to reality via the image; Jimmy's, to truth via the word.

Crake is so unfeeling a character that it is tempting to dismiss him as unbelievable. Sven Birkerts, in his review of the novel for the *New*

York Times, criticizes Atwood for committing what he believes to be typical of the sins of sci-fi: sacrificing personality to premise.[29] This would be a devastating accusation if Atwood intended for Crake to be a real person—someone you could meet today on the streets—but she did not. Crake is a satirical caricature designed to show us something about ourselves in the exaggeration. Crake is a hyped-up version of a unique and new type of narcissist: the technoelite. In her book *Cyberselfish,* Paulina Borsook argues that advanced technological society has produced the digirati, a new would-be ruling elite. Most often white, male, and middle-class, the digirati share more than a love of gadgets and *Star Trek;* they hold fiercely libertarian political views. "Technolibertarianism" is the belief that government should be hands off when it comes to the economy, especially with regard to the internet, technological enterprises, and one's social life.[30] To step into this world with Borsook is to find the culture of Crake and, with it, a convincing prequel to this novel. We are not talking about a couple of isolated computer geeks but a culture whose ideals for society match the conditions necessary for Atwood's Compounds/pleeblands scenario to develop. Technolibertarians are not interested in politics as public service but as a way to enable their work and their freedom to get private funding for whatever utopias they envision.

What makes Borsook's book particularly relevant to *Oryx and Crake* is her assessment of the type of thinking behind technolibertarianism. Borsook describes it as a "kind of scary, psychologically brittle, prepolitical autism. It bespeaks a lack of human connection and a discomfort with the core of what many of us consider it means to be human. It's an inability to reconcile the demands of being individual with the demands of participating in society, which coincides beautifully with a preference for, and glorification of, being the solo commander of one's computer in lieu of any other economically viable behavior."[31] Borsook's use of the word "autism" is metaphorical and trades on her understanding of a complicated condition. Autism is a neurological and developmental disorder with two salient, interconnected features: a lessened ability to be naturally empathetic with others, and a lessened ability to acquire and understand the nuances of language. Autistic children (and adults) are forced to learn how to read nonverbal cues and to learn how to process others' emotions intellectually, since they often find it difficult to feel emotions along with people.

Borsook is not arguing that Technolibertarians are autistic (or that autistic people are Technolibertarians), only that they clearly champion individual genius over social cooperation, numbers over words. Friends sit back to back in a room, interacting with their computers instead of interfacing with others. Social networking, and what it can gain for the networker, supersedes social service. *Oryx and Crake* proves that the elevation of these values is not without consequence. Exaggerated and rewarded by society in the creation of Watson-Crick, the desirable school nicknamed "Asperger's U," these values help to produce a Crake who has no problem contributing to the development of BlyssPluss, a pill designed to increase sexual pleasure while it activates a forced steriliza-tion campaign on those deemed unfit.[32] Crake engineers it one step fur-ther, using the pills to unleash a deadly virus.

Crake may be an exaggeration, but he is not unthinkable. In some ways, he is a typical product of what Christopher Lasch years ago named "the culture of narcissism." Lasch's critique was more than just a 1970s-motivated look at the Me generation: it was an analysis of American culture in which he concluded that some of the traits of pathological narcissism had become mainstream. The narcissist sees others as a mir-ror to the self and thus finds it difficult to connect to the outside world. The three main causes that Lasch identifies read like a description of the near future in *Oryx and Crake*: the breakdown of the family, by which cultural values fail to be adequately passed down; the "dense interper-sonal environment of modern bureaucracy"; and the proliferation of images that have contributed to the creation of a "society of the spec-tacle."[33] Atwood makes more than a passing reference to problems in each of these areas; for example, she develops the familial past of both Jimmy and Crake, who each lost one or more of his parents at a young age. In fact, Crake's father and Jimmy's mother are both victims of CorpseSeCorps, the corporate authoritarian presence that makes it their business to keep the profits coming.

In the 1990 afterword to *The Culture of Narcissism*, Lasch attributed widespread malaise to a cultural tendency toward primary narcissism, or a denial of the feelings of helplessness and dependence on others that all infants eventually experience when they recognize their separation from the mother and the world. Primary narcissists deny the separation and cling to an "illusion of omnipotence" to stave off the anxiety. Thus technology's utopian promise holds a special appeal for them, because it represents a "collective revolt against the limitations of the human condition." In psychological terms, explains Lasch, "the dream of subjugating nature is our culture's regressive solution to the problem of narcissism— regressive because it seeks to restore the primal illusion of omnipotence and refuses to accept limits on our collective self-sufficiency."[34]

By this definition, Crake evinces all the traits of primary narcissism. Crake grows up completely motivated by this dream of the subjugation of nature. He views all human suffering as a problem that can and should be solved by science and technology alone, and thus tries to design a completely new race. The new race, without writing or art, more like animals than people, betrays his narcissistic nostalgic yearning for immediate, womb-like satiation of all desires. Because Crake believes that wars and strife have been caused by "thwarted lust," he designed the Crakers to be more like animals, with mating rituals based on controlled hormone cycles. In all of his plans, reflects Snowman (Jimmy), Crake had "worked out the numbers" so as to eliminate all of the unrequited yearnings of humanity. "It no longer matters who the father of the inevitable child may be, since there's no more property to inherit, no father-son loyalty required for war. Sex is no longer a mysterious rite, viewed with ambivalence or downright loathing, conducted in the dark and inspiring suicides and murders. Now it's more like an athletic demonstration, a free-spirited romp" (*OC,* 165).

REVENGE OF THE G-SPOT

Crake's narcissistic desire is to replace humanity with a species made more in the image of how he sees himself: an autonomous man of science, with no need for others, for art, for language, or for God. With this view of his own self-sufficiency, it is not surprising that he believes that human yearning for God—the ultimate other—is nothing more than misfiring neurons. The view that religious beliefs are primarily physiologically determined is common among the scientific elite today, who often show hostility to theologians or others who argue otherwise.[35] Like these thinkers, Crake blames religion for a number of society's woes, so he tries to eliminate what he calls "the G-spot" in the brain, an operation that is difficult because "take out too much in that area and you got a zombie or a psychopath" (*OC,* 157).

But one of the ways that Atwood critiques Crake's empty utilitarian morality is by locating the spiritual somewhere beyond the control of science. Crake not only fails to eliminate all human beings (Snowman and others survive) but fails to eliminate the spiritual behavior of the Crakers. Needing to know their origins—and needing an answer to the most fundamental of philosophical questions, why is there something rather than nothing?—they create myths based on shreds of stories they have heard from Snowman. Hearing about a creator called "Crake," they begin to worship him. When Snowman, who has allowed them to think of him as a prophet/priest of Crake, leaves them to go on a journey, they build him in effigy and talk to him, chanting his name—"oh man"—in a way that sounds like "Amen." They did so, they explained, to help them to "send out [their] voices" to him (*OC,* 361). The story suggests that spirituality is necessarily linked to interpersonal needs and desires: when the Crakers long for Snowman's company, they create a symbol of him. Language becomes rich and adaptive to absence; stories are told; art is reborn.

In Atwood's postapocalyptic world, language, art, and religious thinking quickly resurge. This is not at all to say that these things are always productive and good—the world of the Crakers is not utopian—but that they are inevitable because they are human. Language, art, and religion are all based on symbolic thinking, and they all presuppose a world of metaphysical desires beyond animal appetites. Even more important, symbolic thinking presupposes the existence of other beings outside of the self, beings to whom we are answerable, as Richard Ford explains: "Language is not first of all about a content to be communicated but is rooted in the orientation to the other, in sincerity and frankness, and in responsibility answerable to the other."[36] Its immediate resurgence here also insists that connection with those beings is primary to existence, not secondary. Narcissistic fantasies to the contrary, it is other beings, and their ethical demands on us, that constitute the self.[37] This is why Snowman is the narrator of the story; his telling reminds the reader that it is the fact of listeners that enable the self to come into being at all.

In naming her tale teller Snowman, Atwood evokes one of the most famous poems about the role of art in the formation of the self: Wallace Stevens's "The Snow Man." It is difficult to determine whether this poem that begins "One must have a mind of winter / . . . And have been cold a long time" is prescriptive or descriptive, whether the speaker warns against a "mind of winter" or simply shows what happens when someone gets it. Either way the poem suggests that the listener's mind must be very cold before he can see a winter landscape without thinking about all the melancholic descriptions of the winter landscape that have come before. The descriptions are artistic creations that have become a part, now, of the world itself. It is only when the listener gets a "mind of winter," and can listen in the snow to the landscape itself, that this listener, "nothing himself, beholds / Nothing that is not there and the nothing that is." But when Atwood's Snowman listens in the stripped landscape, he cannot give himself a mind of winter. He still sees the culture that came before him; he still longs for a future culture; he is desperate for connection. Atwood seems to be saying that Snowman is not nothing himself and that there is something—not nothing—in the landscape. As long as language and a listener exists, there can never be nothing. This is why *Oryx and Crake* ends with the ethical dilemma it does. Snowman finds other humans and has to decide whether to meet up with them or to try to make it on his own. When he considers whether to meet up with them, he thinks in terms of clichéd fragments from the past:

What next? Advance with a strip of bedsheet tied to a stick, waving a white flag? *I come in peace*. But he doesn't have his bedsheet with him.

Or, *I can show you much treasure*. But no, he has nothing to trade with them, nor they with him. Nothing except themselves. They could listen to him, they could hear his tale, he could hear theirs. They at least would understand something of what he's been through. (*OC,* 373–74)

Though Atwood does not reveal Snowman's choice, the need for Snowman to tell his story seems to suggest what route he will take. "They at least would understand something of what he's been through." Without a community to listen to his story, he would have nothing. He would be nothing.

The insistence that the ethical self is constituted in narrative and in community is what makes *Oryx and Crake* more than a sci-fi recapitulation of fears of biotechnology gone amuck. It is instead a prophetic warning: *empower the arts so that they push us beyond the values of production and consumption, efficiency and control.* If we let technique dictate the way we see the meaning of life, love itself will be choked. In a way, the central metaphor of *Oryx and Crake* is the image of vultures eating the dead carcasses of the last four-letter word that the artist Amanda Payne chose to vulturize: "love." Atwood thereby begins to sound a bit like the theologian Hans Urs von Balthasar, who argues that if we view the world as being under the dominion of science, then science and technology "will overpower and suffocate the forces of love within the world" because everything in the world will come to be viewed "solely in terms of power or profit-margin, in which everything that is disinterested and gratuitous and useless is despised, persecuted, and wiped out, and even art is forced to wear the mask and the features of technique." But if we interpret creation in terms of love, argues von Balthasar, then we will understand it. We will make some sense of "the purpose of its existence in general, for which no philosophy can otherwise find a sufficient reason. Why in fact *is* there something rather than nothing?"[38] Though von Balthasar and Atwood would certainly disagree on the answer to this ultimately human question, they would agree on one thing. Someone like Crake should never be left alone to answer it.

CHAPTER 6

I Love Humanity, but I Don't Like You

Walker Percy's The Thanatos Syndrome *and the Soul of Scientism*

> *We ought to embrace the whole human race without exception in a single feeling of love; here there is no distinction between barbarian and Greek, worthy and unworthy, friend and enemy, since all should be contemplated in God, not in themselves. When we turn aside from such contemplation, it is no wonder we become entangled in many errors.*
>
> —John Calvin, *Institutes of the Christian Religion*

There may be no more reliable way to begin to understand the complicated motives behind the biotechnological revolution than to study drugs. Psychopharmaceuticals is a fast-growth industry; the use of antidepressants alone increased by 48 percent between the years 1995 and 2002.[1] Hundreds of thousands of people who have struggled with depression, schizophrenia, ADHD, and other mental health problems have had their lives changed for the better, many of them moving from suicidal incapacity to full functionality.

Rarely has a drug been developed for therapy that does not, in some way, at some point, cross over into being used for enhancement. Not everyone who takes Viagra has erectile dysfunction, not everyone who takes Prozac is clinically depressed, and not everyone who takes Ritalin has ADHD.[2] Even without trying to draw a firm line between therapeutic and enhancement uses, the trends reveal something vital about American culture, the way Americans define themselves, the way we define illness. Before long, it is possible to trace an evolution in attitudes of which the psychopharmaceutical industry is symptomatically illustrative; when drugs are relied on as primary solutions to psychological problems, something has changed in the way a culture identifies its problems and its solutions. As Carl Elliott explains in his book *Better Than Well: American Medicine Meets the American Dream,* people expect much more from medical technology than a cure from disease. Many are turning to it to try to become better or "more authentic" versions of themselves, and quickly discover a consumer culture more than willing to accommodate their wishes by defining a disease to suit. Elliott examines multiple examples of how technological enhancements are marketed with the language of illness and pathology: in each case a "problem" is named, and a "treatment" is prescribed.[3]

This narrative is the one that convinced the twenty-nine-year old Walker Percy, an agnostic Southern doctor, to convert to Catholicism, leave medicine, and become a novelist. The man who desperately wanted the solutions of science soon found that they were not enough; fixing the body did not get at the core issues of the soul. Raised under the specter of suicide, and ill for many years with tuberculosis, he struggled to find meaning for his own life. His illness gave him time to read Kierkegaard, Dostoevsky, and Sartre.[4] His reading led him not only to convert to Catholicism but also to think more generally about why contemporary Americans feel so alienated from themselves. He eventually concluded that the more Americans put their trust in science to solve problems, the more alienated from their true predicament they became. The problem of feeling "lost in the cosmos" was an existential one, not a medical one.[5] So, like Atwood and Huxley, he put his hand to writing about what happens when scientific solutions are writ large—when pill popping becomes *the* way to solve problems rather than a way to help.

Through the character of Bob Comeaux, Percy warns that the problem is not in totalitarianism or in psychopharmaceuticals per se but in having an abstract and thin "Love for Humanity" that leads someone with power to embrace and promote quick fixes and easy enhancements. The problem, which is shared by social engineers and biotechnological libertarians alike, is in being led by scientism to think of individuals as patients who need their problems to be solved, rather than as persons who need to be loved and cared for.[6] Unlike Atwood or Huxley, however, Percy offers an alternative as well as a critique. Through Tom More and Father Smith, Percy suggests that the path to social and personal health must start with love for particular people. Love for particular people, like language, is rich and imperfect, developing along the way and never arriving. To the question of which technologies should be used to help people and why, Percy insists on reframing the question. The question that needs to be asked is, what decisions would a redemptive love for particular people lead us to make?

SCIENTISM AND THE NOVELIST AS DIAGNOSTICIAN

What bothered Percy was that the "man on the street" had begun to believe that something can be called "truth" only when it can be verified by the scientific method. The "typical denizen of the age," Percy writes, believes "as part of the very air he breathes, that natural science has the truth, all the truth, and that the rest—religion, humanism, art—is icing on the cognitive cake; attractive icing, yes, but icing nonetheless, which is to say, noncognitive icing, emotional icing."[7] "Scientism" was the word Percy chose when writing about this misplaced trust in science and its methods. He insisted that his response to science was not simply the response of the classical humanist to the limits of scientific knowledge, namely, that science cannot account for the whole range of human experience, "emotions, art, faith, and so on."[8] When the scientific method is the only way to get at truth, people turn to it to try to answer questions that it does not even claim to be able to answer, such as what it means to be human.[9]

But unlike other critics, Percy was interested less in disproving scientism than he was in diagnosing the cultural malaise caused by it. He was especially struck by the irony that the more science claims to explain every human behavior, the more individual persons feel displaced from themselves and from community. The main reason for this displacement, Percy concluded, is that by definition the scientific method works with types and formulas. While the scientific method tries to come up with laws that explain everything from the behavior of quarks to the behavior of two-year-olds, such theories provide little help in seeing and understanding the individual quirky self. In other words, B. F. Skinner was not only wrong but also a bad novelist. "The peculiar fate of the human being is that he is stuck with the consciousness of himself as a self, as a unique individual, or at least with the possibility of becoming such a self," Percy writes. To the degree that a person perceives himself as a "specimen" of some "biological genotype," that person cannot be himself. And so the "great gap in human knowledge to which science cannot address itself by the very nature of the scientific method is, to paraphrase Kierkegaard, nothing less than this: What it is like to be an individual, to be born, live, and die in the twentieth century."[10]

This "great gap" belongs to the novelist. Percy became a novelist because he believed that neither the meaning of individual lives nor the basis for a common morality can be explained or prescribed by the scientific method. This is a fact to which thinkers as diverse as Daniel Dennett and Jürgen Habermas submit. Though he does not use the term "scientism," Habermas is clearly worried about it, asking, "What will become of . . . persons if they progressively subsume *themselves* under scientific descriptions?" The purely scientific image of the human leads to a "complete desocialization of our self-understanding" and to language that is clearly inadequate for making moral discriminations. In other words, scientific descriptions of human behavior, no matter how thorough, cannot cross over from the realm of the "is" into the realm of the "ought."[11]

Though Percy might have disagreed with Habermas as to how existential and moral questions should be resolved, he turned to fiction, in part, to put the inadequacies of scientific language on display. The novelist, Percy insisted, needs to warn readers that science can solve scientific problems but cannot provide existential answers.[12] That work belongs to the novel. While the scientist looks for answers with rules and theorems, the novelist explores particular human beings through narrative, which is much more personal and a lot less tidy.[13] As Paul Ricoeur illustrates, not only do people experience the narratives of their own lives, but their lives become a shifting mixture of fictional narratives to which they have been exposed: "As for the notion of the narrative unity of a life, it must be seen as an unstable mixture of fabulation and actual experience. It is precisely because of the elusive character of real life that we need the help of fiction to organize life retrospectively, after the fact, prepared to take as provisional and open to revision any figure of emplotment borrowed from fiction or from history."[14] Since no one can escape the role that narrative plays in shaping lives, the question becomes which narratives one should attend to.

Nearly all of Walker Percy's novels respond in some way to the soul (or soullessness) of scientism's narrative. Most of them assert scientism's failure by having for a protagonist a white middle-class man who feels isolated, despairing, and "lost in the cosmos" in spite of the fact that he knows that science can explain more of the secrets of the cosmos than ever before. *The Thanatos Syndrome* also has this kind of protagonist, but it goes one step further in its critique. It is the sequel to Percy's novel *Love in the Ruins,* in which one of his favorite characters, Tom More, first appears. The subtitle of *Love in the Ruins* is *The Adventures of a Bad Catholic at a Time Near the End of the World,* and Tom More is that bad Catholic: a drunk, womanizing psychiatrist. In *Love in the Ruins,* More invents something called an "Ontological Lapsometer," which he uses to attempt to treat the malaise he continues to observe in the human spirit. The comical results of More's lapsometer enable Percy to make fun of the arrogance and pride shared by More and the medical establishment. But *The Thanatos Syndrome* has a different feel to it; it's focused on what More learns as a scientist and person. The novel opens as More, recently released from prison, begins to notice that the people of Feliciana parish have been acting strangely. Upon investigation he discovers that Bob Comeaux, in an effort to improve society, decided to treat the people of their malaises by secretly dumping heavy sodium into the water supply. The project, called Blue Boy, works; crime decreases, teenage pregnancies drop, and people seem to be more satisfied. But in the process the citizens also lose the distinctiveness of their personalities, answering questions more like robots than human beings.

Tom More eventually foils Comeaux's plans, and the region returns to normal. Unlike the end of *Love in the Ruins,* in which More still believes that science can save the world through the lapsometer, by the end of *The Thanatos Syndrome* More has abandoned that scientistic faith.[15]

Through More's abandonment of scientism, Percy illustrates that to truly help others, one must start somewhere else entirely. So he pits Bob Comeaux's sweeping love for humanity against the flawed, halting, and far from perfect love for individual persons that characterizes Father Smith and Tom More. In their clash, Percy illustrates that unattractive and leaky vessels that carry the virtues of hope and love can offer a flawed but viable alternative to an overreliance on technical solutions to human problems.

I LOVE HUMANITY, BUT I DON'T LIKE YOU

Though his character is exaggerated in this satirical novel, Bob Comeaux is meant to be recognized as a child of the age, a scientist who turns all the tools of his trade against the world's problems.[16] He's clearly a scientific naturalist, one who assumes that all aberrations in human behavior will be eventually explained and fixed by science. Writing this novel in 1987, Percy was particularly prescient in that for Comeaux, all the problems with human behavior come down to something in the brain, which, along with the related phenomenon of consciousness, is what most scientists today are calling the final frontier of science. The hypothesis he's operating under, he tells More, is that "at least a segment of the human neocortex and of consciousness itself is not only an aberration of evolution but is also the scourge and curse of life on this earth, the source of wars, insanities, perversions—in short, those very pathologies which are peculiar to *Homo sapiens.* As Vonnegut put it . . . the only trouble with *Homo sapiens* is that parts of our brains are too fucking big."[17] It is not the specific scientific claims that interest Percy here, but Comeaux's assumptions, both about what counts as pathology and about the way to fix it. His speech here has all the rhetorical markers of contemporary scientific naturalism: the scientific terminology, the turn to the brain to explain behavior, the facile assumptions of what counts as pathology, and the references to how science is going to aid our continued evolution as a species. In short, Bob Comeaux is Daniel Dennett with a plan.

Comeaux believes in the Blue Boy project, and like other proponents of the unfettered use of technology as an instrument for good, he rhetorically downplays the fact that it is, inevitably, *his version* of what is good for humanity that is being advanced. The confrontations between More and Comeaux are revealing. When More first asks him about the heavy sodium project, Comeaux emphasizes that he and More have more in common than it appears. "'I'm assuming, Tom,'" he says, "'that we live by the same lights, share certain basic assumptions and goals. . . . Healing the sick, ministering to the suffering, improving the quality of life for the individual regardless of race, creed, or national origin. Right?'" (*TS*, 190). More agrees and then asks what that has to do with putting heavy sodium in the water supply. Comeaux answers that "'one might have asked a similar question fifty years ago: What does it have to do with fluoride in the water supply? And if we'd asked it, we'd have gotten the same sort of flak from the Kluxers and knotheads—as you of all people know'" (*TS*, 191). Comeaux continues, "'What would you say if I gave you a magic wand you could wave over there'—he nods over his shoulder toward Baton Rouge and New Orleans—'and overnight you could reduce crime in the streets by eighty-five percent?'" (*TS*, 191). He goes on to point out that not only is crime down, but so is teenage suicide, wife battering, depression, and eventually even AIDS, because the sodium supply has reduced the incidents of homosexual practice and intravenous drug use.

Of course, this technology does not exist outside the world of the novel, but Comeaux's rhetorical approach to its use is present enough. In his book *Liberation Biology*, for example, Ronald Bailey tries to paint those who urge caution in the face of the biotechnological revolution as fearful of making positive changes. Bailey wants individuals to have the right to liberate themselves and future generations from everything that makes our human life a struggle: "In the twenty-first century, liberation biology is the earthly quest to overcome the physical and mental limitations imposed on us by nature, enabling us to flourish as never before." The world toward which the biotechnological revolution is aiming is "one in which more and more individuals can exercise enhanced intellectual, creative, and physical capacities while being liberated from the immemorial curses of disease, disability, and early death."[18] Through characters like Bob Comeaux, Percy's satire shows how tempting the idea of having a magic wand to wave over New Orleans and, indeed, the whole human race will always be for leaders. Comeaux and Bailey simply rhetorically trade on areas of complete agreement in society. Who could disagree that less crime is a good thing?

Additionally, both Comeaux and Bailey assume that nature has dealt humanity a bum hand and that it is time to take charge of the problem ourselves. Bailey, like Comeaux, argues that his vision, therefore, respects the sanctity of human life even more than does the likes of Leon Kass or Tom More. Bailey argues that "who really has a higher regard for the sanctity of human life—those who, like Kass and Bill McKibben, fatalistically counsel us to live with the often bum hands that nature deals us, or those who want to use genetic technologies to ameliorate the ills that have afflicted humanity since time immemorial? Respecting the sanctity of human life doesn't require that we take whatever random horrors nature dishes out."[19] In *The Thanatos Syndrome,* Comeaux is bothered by these "random horrors" and challenges More to "name one thing we're doing wrong." When More answers with the "technicality of civil rights" and says that "you're assaulting the cortex of an individual without the knowledge or consent of the assaultee," Comeaux gets riled up and insists "let me tell you about assault and who's assaulted!" (*TS,* 193). He points in the direction of Angola Prison, where there are "ten thousand murderers, rapists, armed robbers, society's assholes, who would as soon kill you as spit on you. That's where the assault comes from" (*TS,* 194).

It is through Comeaux's angry outburst here that Percy puts a fine point on it. It is exactly how Comeaux and Bailey and other proponents of biotechnological solutions see "society's assholes" that is at issue here. Percy shows that Bob Comeaux is frustrated with a problem that he thinks comes not from him but from "out there." His frustration, however, is the real problem. It comes from arrogant impatience. Comeaux's rant against "society's assholes" is no different from Bailey's energetic desire to eliminate the "random horrors" in society: both try to redefine love away from the virtues inherent in learning to love real humans and to make it about doing everything we can to make those other humans more loveable, even to make them worthy of having a life to begin with.[20]

This kind of contempt, often masked as compassion, has a long history. As Charles Rubin recently explained in his essay in the new collection from the Presidential Commission on Bioethics, the most radical advocates of remaking human nature—namely, groups called transhumanists, extropians, or singulatarians—are simply following the lead of Francis Bacon and René Descartes in "believing themselves to be the true defenders of human dignity against all the indignities imposed on us by the naturally given: disease, deprivation, decay, and death."[21] But their search for this "incomprehensible human future" in effect deprives the term "dignity" of determinate moral meaning. As Rubin explains, when dignity takes its definition from future versions of the person, it is usually accompanied by contempt toward actual persons. The dignity that these Promethean dreamers are looking for does not characterize real persons and relationships but is based on a negation of their characteristics.[22] Bob Comeaux clearly fits into this camp.

Percy does not stop with a critique of Comeaux's characters and motives; he also critiques the results of his plan. Percy looks in on the actual people who have been "helped" by the technology. *The Thanatos Syndrome* is a work of fiction, so of course the technology is simplified and Comeaux's Blue Boy pilot works. Crime drops in the region, and there are fewer and fewer complaints of depression. Additionally, beams Comeaux, "L.S.U. has not lost a football game in three years, has not had a point scored against them, and get this . . . has not given up a single first down this season" (*TS*, 195). Percy never lets the reader forget that the novel is a satire, but he does an effective job outlining what might happen to humanity if all personal struggles are defined as medical problems to be solved. Percy emphasizes that blanket technological solutions that target certain parts of the brain in order to control human behavior have a downside that is a clear flip of the upside: a one-size-fits-all technological solution might create a one-size-fits-all humanity, too. The novel's opening medical intrigue is built around how Tom More first notices the syndrome: his patients stop acting like themselves and start acting more like each other. They are evened out, smoothed into a kind of compliance. One patient, Mickey LaFaye, has been in therapy with him for years. She has always had a certain mannerism in which she would duck her head, touch her neck, and look at him out of the corner of her eyes. That mannerism disappears and is replaced by a mild and unfocused gaze. Even more significant, More notices that what used to characterize her personality—how she would sit in his office and yearn in a manner characteristic of Andrew Wyeth's Christina—has been changed into a personality more reminiscent of the "satisfied Duchess of Alba" (*TS*, 6).

What More begins to notice about his clients is that while they don't exhibit some of their former problems, they have lost something that was essential to them, something about them as particular persons. More notices it because the clients begin to use language out of context; they begin to answer questions more as a computer might. So Percy has More tell the reader that "they utter short two-word sentences. They remind me of the chimp Lana, who would happily answer any question any time with a sign or two to get her banana." But, More continues, you (the reader) would never use language that way: "You'd want to know why I wanted to know. You'd want to relate the question to your—self" (*TS*, 22). Of course, Percy is not making any scientific claims here, as if enhancement technologies that target the brain are going to necessarily cause this kind of decontextualization of the self. Instead, he emphasizes the link between an individual's self and his or her use of language, that symbol-making activity that Percy repeatedly argues makes people human.[23] Percy's concern is that medical solutions that think about the human mind as one would think about a computer—in terms of input/output—might encourage people to behave more like one.

LOVE YOUR NEIGHBOR AS YOU DO YOUR (MESSY/ HUMAN/PARTICULAR) SELF

Percy locates his alternative vision in two of the most unlikely characters: Tom More, the womanizing, ex-con, and often drunk psychiatrist, and Father Smith, the loony and ex-drunk priest.[24] The name "Tom More" of course references the utopian and dystopian tradition in literature, but Percy also chooses it because Thomas More is his ideal, an "English Catholic—a peculiar breed nowadays—who wore his faith with grace, merriment, and a certain wryness."[25] More achieved what Percy aims for: he mocked the surrounding culture, but without disrespect.

Though the character Tom More is a "lapsed Catholic," his Catholic habit of love resists Comeaux's vision. At the end of the novel, More and Comeaux come to a gentleman's agreement: More will not turn Comeaux over to the justice department if Comeaux, in turn, shuts down the Blue Boy pilot and diverts the ill-gotten federal funds back to the parish hospice. He does so and then leaves the parish. But before he leaves, he tries to get More to agree with him that there is no essential difference in their goals, just their tactics. While More is good at helping people one on one, Comeaux claims he is better at the "ultimate goal" of "the greatest good, the highest quality of life for the greatest number" (*TS,* 346). Comeaux insists that More cannot "argue with the proposition that in the end there is no reason to allow a single child to suffer needlessly, a single old person to linger in pain, a single retard to soil himself for fifty years, suffer humiliation, and wreck his family" (*TS,* 346). He insists that they are after the same thing and that More cannot name one reason why what he is doing is wrong. The only difference between them, he tells More, is that "you're in good taste and I'm not. You have style and know how to act, and I don't" (*TS,* 347). Before More has a chance to respond, Comeaux walks off.

This scene is vintage Percy. It is a prescient insight into the soul of scientism and its most common ethical partner, utilitarianism. It is bad enough that Comeaux tries to pass off their differences in the Southern genteel way, as mere differences in style. But what is worse here is the Hegelian energy in Comeaux's rhetoric: the effort to explain everything and to roll all differences up into a space of ideal agreement. This scene is more than a mere illustration of Comeaux's arrogance (though it is that). It represents the central and increasing difference between Comeaux and More. What motivated Percy as a novelist was the need to rebut the hubris of any kind of philosophical, ethical, or scientific system that subsumes the particularity of the individual. This is why Percy so often reiterated Kierkegaard's critique of Hegel. The closest that Percy would come to defining the self is to say that it is what is left-over when defining is done: "No matter how powerful the theory, whether psychological or political, one's self is always a leftover. Indeed, the self may be defined as that portion of the person which cannot be encompassed by theory, not even a theory of the self."[26] People are never simple, and they need much more than theory to deal with questions about how best to live with themselves and others.

Percy's belief in the unsubsumable particularity of persons leads him to all his wacky characters, but especially to Father Smith. Like the crazy, backwoods, fundamentalist characters of Flannery O'Connor's stories, Smith is Percy's all-too-human prophet, the one whose words seem completely deranged but must not be ignored. Father Smith's first prophetic action is to isolate himself in a fire tower and refuse to preach because "words no longer signify." One of the funniest scenes in the book is when More goes up to the tower to try to persuade Smith to return to his parish duties. More quickly notices that Smith has gone over the edge. "Father Smith has gone batty," More tells us, "but batty in a way I recognize. He belongs to that category of nut who can do his job competently enough, quite well in fact, but given one minute of free time latches on to an obsession like a tongue seeking a sore tooth" (*TS*, 120). More has no choice but to hear Smith's obsessions; he is literally trapped by Smith's foot that is placed on top of the only door that leads up into his tower room. Smith launches into a diatribe about how words have been deprived of their meaning, so that there must be a depriver. There is one word sign, he insists, that has not been deprived of its meaning, and he tries to demonstrate it to More by playing a word association game. Smith asks More to free associate around the terms he names: "Irish," "blacks," and "Jews." Since More names specific people only when he hears "Jews," the game supposedly proves that the Jews are the one sign that cannot be subsumed into abstraction. Smith exults that they "were the original chosen people of God, a tribe of people who are still here, they are a sign of God's presence which cannot be evacuated" (*TS*, 123). In spite of More's increasing efforts to escape the conversation, Smith eventually links this idea to the effort in the Holocaust to kill all Jews and then tells More that "you are an able psychiatrist, on the whole a decent, generous, humanitarian person in the abstract sense of the word. You know what is going to happen to you?" (*TS*, 127). Smith then tells him that since his generation of doctors is the first to permit euthanasia and abortion, then they will also end up killing Jews. "Somehow I knew he was going to say this," reflects More, and now he's really desperate to get down from the fire tower.

Although Smith's logic is uneven—and the reader is supposed to recognize that fact—all of the core ideas of this diatribe belong to Percy. The Christian response to Comeaux-like efforts to abstract, name, and solve humanity's problems is the scandal of particularity. Part of what is scandalous about Judeo-Christian teaching is that it insists that the Jews were a particular tribe that was chosen by God in space and time and that Jesus was a particular Jew in a particular time as well. "Christianity is doubly offensive because it claims not only this but also that God became one man, He and no other," Percy explains. "One cannot imagine any statement more offensive to the present-day scientific set of mind."[27] The Christian idea of the sanctity of every human person rests inexorably on the fact of God becoming an individual person himself. Problems come when humanity is abstracted, when the individual is lost in the name of doing good for humanity. In this satiric novel, Percy tries to get the reader to think about it by exaggerating the ideas in Smith's loony prophecies. Like a lot of fools in literature, Smith spouts wisdom. So before More can leave, Smith gives him his "final word," in which he claims that while lovers of "Mankind in the abstract" who write poetry, like Walt Whitman, are harmless and theorists of "Mankind" like Rousseau and Skinner are also harmless, "if you put the two together, a lover of Mankind and a theorist of Mankind, what you've got now is Robespierre or Stalin or Hitler and the Terror, and millions dead for the good of Mankind" (*TS*, 129).

As extreme as Smith gets, his critique shares much with Christian personalism.[28] Love for humanity in the abstract cultivates, as Flannery O'Connor first wrote and Father Smith repeats, the "tenderness that leads to the gas chamber."[29] The problems multiply when a superficially abstract love for humankind meets a reductively abstract theory of what is best for humankind. What bothered Percy so much about contemporary theories of anthropology, especially under the rule of scientific materialism, is that what has been traditionally viewed as human distinctiveness (positive differences in people) or as sin (willful disobedience to moral law) could, willy-nilly, be redescribed as pathology. This move permits and even encourages an earnest reformer to love humanity by treating what he merely dislikes as a disease or a problem. Eugenics is the ultimate example, and there is nothing inherent in a scientific naturalist theory of humankind that could be used to argue against it. The only thing left to argue over is whose version of human perfection we should strive for and how best to strive for it.

Induced by the Incarnation and the doctrine of the *Imago Dei,* Christian personalists insist that love for persons must be love for particular persons whose distinctiveness comes from God, not from random errors of nature. Love then requires a separation of the person (as created in the image of God) from her behavior (good or bad). The point here is one that has been made by Christians for centuries: when one believes, or habitually acts as if one believes, that all persons are made in the image of God and that this God became a man to redeem the tarnishing of that image, love for persons will necessarily look different. Love will truly celebrate differences and be patient when it comes to sin, frailty, and suffering. Love will be saturated in grace and therefore slow to name anyone "society's assholes." In his *Works of Love,* Kierkegaard argued that what he called "small-mindedness," which is the inability to love human distinctiveness, comes from the failure to see one's own distinctiveness as a gift of God. "To have distinctiveness is to believe in the distinctiveness of everyone else, because distinctiveness is not mine but is God's gift by which he gives being to me, and he indeed gives to all, gives being to all."[30] Existence itself is the primary gift and sign of God's love. As Erasmo Leiva-Merikakis puts it, "that anything at all, not only flowers but also you and I, should nevertheless exist: *this* is the fundamental truth throbbing with the goodness and beauty and glory and love of God."[31] Since being is a gift of God's love, human love should also start with a simple affirmation of being. This is why Leiva-Merikakis points out that the main pre-Christian use of the Greek word *agape* (love) was to designate the action of welcoming someone in one's home. Part of what is then implied in the use of *agape* to describe God's love for humankind is that God's love constitutes the ultimate example of welcoming the other.

In his celebration of the goodness of being and the distinctiveness of each human person, Percy joins an august tradition of Catholic novelists, including Francois Mauriac, Georges Bernanos, and Flannery O'Connor. What sets Percy apart from these is his insistence that this distinctiveness is primarily revealed in language and that language is inherently dependent on personal relationship. The human is a symbol-mongering wayfarer, and "symbolization is of its very essence an inter-subjectivity." Symbolization is an act of affirming being that is dependent upon a Thou that corresponds with the I and encounters a real world together. Percy writes that, "the act of symbolization is an affirmation: Yes, this is water! My excitement derives from the discovery that it is there for you and me and that it is the same thing for you and me. . . . *Symbolization presupposes a triad of existents: I, the object, you.*"[32] Communication is inseparable from the recognition that other people inhabit the same world that I do and that their existence is not incidental to mine.

That there is minimally a real and personal triad behind every act in language is an idea that comes directly from Percy's neo-Thomist sacramentalism.[33] It also illustrates Percy's affinity with the work of some contemporary theologians who draw on the Trinity to suggest, among other things, the transformation in ethics when reasoning about what it means to be human starts with relationship rather than with individual consciousness.[34] Jean Zizioulas, drawing on patristic teaching, insists that personhood as a historical concept would not be possible without the personal and loving existence of God in the Trinity. "True being" he argues, can only come from the person who "freely affirms his being, his identity, by means of an event of communion with other persons."[35] It is in communion that the image of God—the *Imago Dei*—is manifest. Without communion, there are no persons.

These ideas represent a stridently countercultural perspective when allowed to engage with current trends in bioethics. What theologians like Colin Gunton, Zizioulas, Christoph Schwöbel, and others share with Percy is the conviction that when a Trinity of persons in loving relationship directs the way we think about humanity, it necessarily resists the definition of the person as construed by the Cartesian *cogito ergo sum* ("I think; therefore, I am"), wherein the self is birthed by the self. Whether it is ontologically true or not, it is potentially ethically devastating to think of oneself as an autonomous individual rather than as a part of (or even a product of), a community of persons in which loving relationship is the ultimate goal.

Robert Spaemann develops his ethical inquiry along these lines. In his book *Persons: The Difference between Someone and Something,* he insists that we are persons not because we cogitate within the *cogito,* but by virtue of our having been born into a community of other persons, persons who recognize us.[36] Like Paul Ricoeur, Spaemann insists that recognizing the irreducible reality of the other person is the true beginning point of love and justice. I will quote the following passage at length because of its clear connection to Percy's ideas:

> The elementary form of such absolute encounter with reality is the intersection of the other's gaze with mine. I find myself looked at. And if the other's gaze does not objectify me, inspect me, evaluate me, or merely crave for me, but reciprocates my own, there is constituted in the experience of both what we call "personal existence." "Persons" exist only in the plural. It is true that the gaze of the other may in principle be simulated, for the other is never presented to us in the compelling immediacy of pure phenomenon. It is a free decision to treat the other as a real self, not a simulation. What that decision essentially consists in is a refusal to obey the innate tendency of all living things to overpower others. Positively expressed, we may call it "letting-be." Letting-be is the act of transcendence, the distinctive hallmark of personality. Persons are beings for whom the self-being of another is real, and whose own self has become real to another.[37]

Walker Percy's own ideas about what is at stake in our view of ourselves and the other map onto Spaemann's quite readily. *The Thanatos Syn-*

drome doesn't prove the ontological point that persons are constituted by relationship with one another, but it does insist on the ethical point that it is how the lover of humanity views humanity will dictate the quality of his love and its outcome. In other words, one of Bob Comeaux's primary problems is that he does not see himself in relationship with the people he is trying to help. He does not see himself as seen by the other. He himself is depersonalized as he depersonalizes the people he is trying to help. Even when he asks Tom More to give him feedback on his project, he does so in a way that sets the parameters of the critique in general, social terms, not specific, individual ones. "'Tell me honestly. Don't pull punches,'" Comeaux insists. "'Has anything you've heard in the last few minutes about the behavioral effects of the sodium additive struck you as socially undesirable?'" (*TS*, 201). When the question is put that way, More can only mumble that he will have to think about it. The problem is that this question cannot be answered by a simple yes or no.

As I have mentioned, the novel does not end with Bob Comeaux's failures but with Tom More and his actions that, though they are not perfect, comprise a genuine alternative. At the end of the novel, things begin to return to normal in the parish, LSU starts losing football games, and More returns to his vocation as a psychiatrist. He talks again about the lives and halting progress of each of his patients. One of the patients who finally returns to see him is Mickey LaFaye, the first patient whose behavior had led Tom More to investigate the syndrome. Mickey has nearly returned to her former self; she is now "no Duchess of Alba" but "almost Christina again" (*TS*, 370). He asks her to sit in his room so that they can see each other, and it is in this face-to-face exchange that she begins to explore a recent anxiety attack. By question and answer he slowly draws her out, gets her to talk about dreams she has had, encourages her to try to figure out what the various aspects of her dream might mean to her, and so on. The scene provides a direct contrast to the methods employed by Bob Comeaux in both the Blue Boy pilot and his rhetorical bullying.[38] Mickey asks More if he knows who the stranger in her dream might be, and their dialogue continues (*TS*, 371–72):

"Who do you think he is?"

"I think the stranger is a part of myself."

"I see."

"I am trying to tell myself something. I mean a part of me I don't really know, yet the deepest part of me, is trying to—"

"Yes?"

"Could I talk about it?"

"Yes."

She falls silent, but her eyes are softer, livelier, are searching mine as if I were the mirror of her very self. She lets go of her hand. She almost smiles. She ducks her head and touches the nape of her neck as she used to.

"Well?" I say.

She opens her mouth to speak.

Well well well.

This scene is a fitting conclusion to the book for a number of reasons. Mickey has returned to the mannerisms and behaviors that constitute the distinctiveness of her personality, a distinctiveness that had been nearly destroyed by the sodium treatments. She is ducking her head and touching the nape of her neck, and she is yearning for something she cannot quite name. There's a sadness here, but because it belongs to Mickey uniquely, it can become the starting point for self-discovery. Her appearance in More's office illustrates Percy's conviction that the self must go through language to discover itself, and it is a process that is slow and halting. It is a process that will only begin if the anxiety provokes it; in other words, the forced happiness of the cortical solution of the Blue Boy project blocked the very avenue of her self-discovery when it smoothed her out. In an interview, Percy explained that he felt a lot of sympathy with the existentialists' project because it revealed, as Kierkegaard had demonstrated philosophically, how anxiety is a warning from the authentic self to the inauthentic self, a tip off to the existence of despair that does not recognize itself as despair.

Another way to put this is to say that the first Christian virtue exercised in their encounter that was effectively eliminated by Comeaux's plans is hope. More has genuine hope in Mickey's progress because she is able, again, to learn from her anxiety. "She is about to listen to herself tell herself something," Percy said in an interview.[39] It is a completely open-ended conclusion: Mickey opens her mouth to speak, and she is on the road to a recovery that will never be complete. A person starts out as a symbol-mongering wayfarer and will end up there, too, but she *can* get a little further down the road. For Percy, the contemporary technological effort to solve all of humankind's problems is flawed in part because it locates its hope in an end product that the person was never meant to be.[40]

But the main reason Percy ends the book with this scene is because it highlights More's love for Mickey as neighbor, a love that, while imperfect, is neither sentimental nor abstract. Tom More still has problems, but that does not mean that this relationship will fail to be productive of real hope and wellness. The two people are face to face, and Mickey is searching More's eyes as if he were "the mirror of her very self," which, in fact, he is. She can never see herself completely by herself. The process requires dialogue, not the bare interior monologue of the *cogito* cogitating with itself. The scene reveals a speaker and a listener in a triadic relationship with the subject of their discourse: the becoming self. There is no triumph here, no completion, no answers.[41] There is dialogue, discovery, and help, but there is also "letting-be." The double meaning of the last three words—More's unspoken thoughts—says it all. Affirming both his satisfaction at Mickey's opening her mouth to speak and declaring the promise of where this will arrive eventually, More thinks to himself, "Well well well." Mickey comes full circle, and Percy's prophetic warning is complete. Percy is not against psychopharmaceuticals; there is nothing in his work to suggest that a person who is depressed should avoid seeking every kind of help. But when it comes to the existential condition of every woman and every man, nothing can replace the metaphysical work a soul must do inside a loving community to find true wellness.

PART
IV

From
Posthuman
Individuals to

Human Persons

CHAPTER 7

Technology, Contingency, and Grace
Raymond Carver's "A Small, Good Thing"

> *In each lived moment of our waking and sleeping, we are technological civilisation.*
>
> —George Grant, *Technology and Justice*

Raymond Carver is not typically the writer we talk about when we talk about technology. When we want to explore our technological society, we usually turn to speculative fiction, stories in which megalomaniacal scientists accidentally kill all humans, or stories in which machines become self-aware and go on a rampage. But to limit our discussion to these stories is to limit our understanding of just how deep the assumptions of our technological society run and how many are its ramifications. Although Carver did not write about technology per se, his story "A Small, Good Thing" grapples with it on a more fundamental level. The story reveals how completely and devastatingly our technological society has embraced one of its primary goals: to eliminate contingency. As C. S. Lewis has argued, the goal of modern science applied through technique is to "subdue reality to the wishes of men," and its most ambitious proponents believe it possible to one day completely control nature and spare humans from any suffering.[1] Carver's story challenges the ascendancy of this goal by showing how it has isolated us from our neighbors and thinned us morally. The story dares to suggest that contingency, even radical contingency, is a small, good thing in an age such as ours. Properly understood and received, contingency is one of the few ways that people who live in an advanced technological society can learn to receive grace.

Carver published his final version of "A Small, Good Thing" in the collection *Cathedral* in 1983.[2] It is a gut-wrenching tale of a woman who orders a birthday cake for her only son's eighth birthday party, only to have him get hit by a car, enter a coma, and die three days later. The conflict in the story comes from the fact that the baker she had ordered the cake from, angry about the fact that no one had picked it up, continues to call the family, ignorant of the boy's death. The parents had, of course, forgotten all about the cake, so the unidentified caller who keeps gruffly saying, "Have you forgotten about Scotty?," seems to them to be some kind of psychopathic maniac. Eventually they remember the order and go to the bakery to confront the baker. After they express their anger, he apologizes and invites them to sit with him. They eat fresh bread together and share stories, talking into the dawn.

This brief plot summary does not do the story justice, of course. One of its strengths is that with a few sparse sentences, Carver is able to render a scenario we all recognize: in America, consumer capitalism has largely turned neighbors into nodes of impersonal economic exchange. Carver claimed that a story always came to him by way of its first sentence—a sentence that he rarely changed—and the first sentence here explains a lot: "Saturday afternoon she drove to the bakery in the shopping center" (SGT, 59).[3] It is Saturday in the suburbs, and people drive to nameless places in order to consume goods. Ann Weiss, the child's mother, attempts to be friendly, but the baker is all business; he is abrupt, and "there were no pleasantries between them, just the minimum exchange of words, the necessary information" (SGT, 60). Though the baker plays a major role in the story, Carver never gives him a name beyond "the baker," and it is clear that for quite some time he has thought of himself only as what he does. The baker's later description of how he repeats his days "with the ovens endlessly full and endlessly empty" (SGT, 89) makes him exemplary of what Hannah Arendt calls the ascendancy of *"animal laborans,"* the laborer. In her book *The Human Condition,* Arendt worries that human dignity has been compromised under the modern reign of *techne,* in which people are reduced to the roles they perform. Arendt's concerns are laid out in this story: the baker has no public visibility or sphere of action outside of his labor and its exchange. His abruptness with his customer despite her polite efforts at small talk is glaringly typical of most encounters in contemporary American society. Words that could be used to connect people are reduced to either an exchange of pleasantries or information, nothing more.

Although technology is not the cause of this loss of a meaningful public communication, it is not neutral in it either. While twenty-first-century people have grown accustomed to thinking of technology in terms of computers and iPods, Carver is dealing with technological advances that are far more pervasive and, therefore, more invisible in their effects. It has been well documented, for example, that privileging the American auto industry has contributed to the fragmentation of traditional neighborhoods and communities and that the telephone increasingly has replaced face-to-face conversation.[4] The action in Carver's story highlights both these facts: the driver who hits Scotty with his car is a stranger, and the baker's anonymous use of the telephone causes the main conflict. By the 1980s, both the automobile and the telephone had become culturally invisible, so when Carver centers his story around a situation that would not have happened without them, their life-changing significance sharpens into focus.

Another technological device that Americans were completely enculturated to by the 1980s is the television. But in Carver's story, as in Don DeLillo's 1984 novel *White Noise,* the television becomes one of the main reasons why language that goes beyond the banal, the functional, or the merely imitative is harder to find. After the death of her son, Ann notes with horror that she cannot seem to speak any words except those that had been "used on TV shows where people were stunned by violent or sudden deaths" (SGT, 81). Even Scotty's doctor is reduced to the clichéd expressions we use when we do not know what to say: "I can't tell you how badly I feel. I'm so very sorry, I can't tell you" (SGT, 80).

If Carver's scenario seems unremarkable, that is precisely the point. Albert Borgmann's main concern in his many treatments of the issue is that technology has become an unseen yet firmly established pattern that structures all human relationships. It is in the quotidian, "the dailiness of modern life," that technology has become most influential.[5] The car, the telephone, and the television have all contributed to the fact that neighbors have been transformed into consumers who are invisible to each other, made shallow, and defaced. But from a Christian perspective, argues Borgmann, the worst thing about the culture of technology is deeper and more fundamental. The worst thing about it is its inherent goal to eliminate contingency. Contingency, he explains, is a "technical term for what lies beyond prediction and control."[6] To eliminate contingency is to be able to predict and control everything.[7] Borgmann explains that this effort to eliminate contingency erodes the conditions necessary for Christianity to prosper because Christian belief begins with receptivity to grace, which is, by definition, "always undeserved and often unforethinkable."[8] Grace is a gift people best learn to receive in situations that highlight their fundamental lack of control. As a result, any culture set on the elimination of contingency makes it harder to experience grace, harder to write about its appearance, and harder to learn anything from it. Put simply, a culture that makes it difficult to receive grace also makes it difficult to receive Christ.[9]

Raymond Carver is just one of many writers who reveals how contingency punctures our illusions of control and leaves a real opportunity for grace in the process. Martha Nussbaum explains that the contingent is well-charted territory for novelists; the very structure of the novel has a built-in "emphasis on the significance, for human life, of what simply happens, of surprise, of reversal." Although Nussbaum does not discuss the concept of grace per se, the Aristotelian conception of learning she believes that novelists possess is consistent with Christianity because they start in the same place. Both insist that people learn the most (and the most important things) from events that are outside of human control. Novelists, argues Nussbaum, cultivate the reader's perception and responsiveness by sharing stories in which the characters' attempts at achieving the good life are challenged by contingency. It is too facile to say that a reader learns a "moral" by reading a story. What a reader learns is how to think about contingency when it appears in his or her own life. In other words, the reader learns to think of contingency as a possible avenue for grace. As Nussbaum explains, the ability to read a situation is an active task that "is not a technique; one learns it by guidance rather than by a formula."[10]

Just how much "moral guidance" Raymond Carver can be interpreted as providing his readers is the subject of critical dispute.[11] There is no doubt that Carver's recovery from alcoholism in 1978 gave him new insight into grace and gratitude, insight that transformed his writing in ways that were extremely significant to him. "A Small, Good Thing" is exemplary of the change. But how radically and quickly his fiction changed has been obscured for years by the amount of editing that his first postalcoholism book, *What We Talk About When We Talk About Love* (1981), underwent. Although the earliest published version of the story, entitled "The Bath," abruptly and bleakly ends right after Scotty's death, recent scholarship confirms that the story we now have as "A Small, Good Thing" was much closer to the original, which Carver composed in 1980.[12] "The Bath" is the product of two stages of radical cuts made by Carver's editor at Knopf, Gordon Lish. Lish cut Carver's original story by 78 percent, prompting a desperate letter from Carver in which he told Lish that he wanted to pull out of the project because the stories as he wrote them were "so intimately hooked up with my getting well, recovering, gaining back some little self-esteem and feeling of worth as a writer and a human being."[13]

One can only speculate about the connection between these stories and Carver's recovery, but it seems to have something to do with this new understanding of grace. The most disturbing thing about Lish's editing job was how he later reported that what drew him to Carver's work was its "peculiar bleakness," so he cut the stories in order to foreground that.[14] To be sure, because grace is not the necessary outcome of a bad experience, neither version of the story is untrue. But Carver's version differs as radically from Lish's version as Carver's life differed from

Lish's: Carver barely escaped death from alcoholism and began to interpret the experience of being an out-of-control alcoholic as the thing that
saved his life, the thing that later enabled him to see the remainder of it
as "gravy."[15] In other words, unlike "The Bath," "A Small, Good Thing"
insists that a certain amount of trouble is required for grace to be made
visible. The story supports Borgmann's claim that "trouble is often the
twin of grace, and if one cannot prosper, neither can the other."[16]

Borgmann's use of the word "prosper" here requires some qualifica-
tion. He is very careful to explain that Christianity has always been
about helping to alleviate the suffering of others and that it is not a
Christian answer to go about looking for suffering. Likewise, my point
is not to argue that Carver, playing God, "uses" Scotty's death to teach
Ann Weiss and us a lesson. Carver's story does not attempt to blame or
to vindicate God for creating a world that allows for accidents, addictive
tendencies, evil behavior, and death. Instead, "A Small, Good Thing"
suggests that the primary tragedy of contemporary life is the fact that
people who live in a technological society expect to exert so much control over their isolated "domestic cocoons" that grace can be revealed
only through radical contingency.

The typical grace-precluding resilience of the domestic cocoon is
why Carver takes care to weave into the story an encounter between
Ann and another family whose son was in the hospital and who also
dies. The particulars of the encounter are important: Ann is looking for
the elevator when she stumbles into the waiting room where they are.
Had she found the elevator, that technology would have carried her to
her destination faster, keeping her from conversing with the family. By
making the family African American, Carver might be suggesting that
Ann's heavy burden caused her to come out of her comfort zone and
converse with people she normally might not have. As she listens to
their story and hears that they, too, have been reduced to hoping and
praying, she wants to reach out to them. When she sees the mother's
lips moving silently, "she had an urge to ask what those words were. She
wanted to talk more with these people who were in the same kind of
waiting she was in. She was afraid, and they were afraid. They had that
in common" (SGT, 74).

Carver's oeuvre is full of examples of stories in which language offers only the barest of connections with people and is continually seen as an inadequate vessel of the deepest of concepts, such as in the story "What We Talk About When We Talk About Love." But this scene offers some kind of new beginning for Ann, one in which Ann can imagine a different kind of responsibility outside of that provided by her family. The experience of radical contingency paved the way for both families to give grace to each other as persons.[17] In his book *Persons: The Difference between 'Someone' and 'Something,'* Robert Spaemann explains that pain is one of those conditions that drives us to wonder how our experience of it differs from others' experiences. To find out, we must work together "with the aid of words."[18] And when we begin to talk, we begin to recognize others as persons, which is to begin to express respect. Persons speak *to* other persons, not only about them.

Reaching out to one another as persons is, of course, central to love. It is also nothing less than the starting point of justice. Spaemann puts it simply: "all obligation begins with noticing persons."[19] The story pushes us to ask a very uncomfortable question: Does it take such an extreme contingency for Ann and for us to step out of ourselves and move toward another? If Nussbaum is correct that "morality is a response to the fact of suffering," at what point have people of the Western world so eliminated contingencies (or think we have), so eliminated suffering (or think we have), that we can no longer empathize with others and therefore can no longer work truly for justice? Would we even be able to recognize the loss?

Experiencing contingency does not guarantee that compassion will be generated and justice will be served, however. Ann, like everyone who lives in an advanced technological society, is not accustomed to losing even a little control, and the accidental death of her son represents a situation wildly out of her control. Because Scotty's death is a contingency that cannot be justified by any lesson it might teach, the tension in Carver's story resides in the possibility of Ann's descent into despair, of the event not serving as a catalyst for deeper empathy for others. Ann's loss has been substantial; she's angry, and when she figures out that it is the baker who has been calling, her anger finally has a target: "That bastard. I'd like to kill him," she says, "I'd like to shoot him and watch him kick" (SGT, 83). Although this violent response is clearly out of proportion with the offense the baker has committed, the reader shares Ann's anger and desire for revenge. We are angry at the baker, too, because we are angry about Ann's loss and do not know whom to blame. Although the baker did not mean any evil, Ann experiences his actions as evil in the context of this loss. In his essay "Evil in Person," Jean-Luc Marion argues that the pain of experiencing evil causes one to try to find and suppress its cause. But often the cause is misidentified, because the immediate concern is to find "any interlocutor whatsoever" on whom to unload the hurt.[20] This driving need explains Ann's impulse for revenge and also our desire to participate in it.[21]

The "domestic cocoon" scenario that Carver created intensifies this impulsive reaction. Ann has no one to talk to except her husband, who has also experienced this loss and is trying to deal with it in his own way. So as she heads out to punish the baker with her anger, things could get even worse for her. The story is beginning to look like many contemporary treatments of how the anonymity of our lives, permitted and reinforced by technology, leads to action and reaction without genuine human connection.[22] Ironically, Ann is about to deprive herself of what she really needs at this time, which is to begin to salve her loss through a deep communion with others.

Borgmann explains how living in an advanced technological society makes the transformation from anger to forgiveness and grace even more difficult. If an individual's goal is to control nature and prevent all contingencies, when accidents do happen, the result is most likely to be frustration and uncomprehending anger. Ann simply has no resources to think of her suffering in any way except to protest that it just isn't fair and that someone should have to pay.[23] Ann keeps the information about her son's death to herself and begins to accuse the baker, and the tension mounts. The baker is even tapping his rolling pin in his hands when Ann finally blurts out that her son is dead. The reader's heart is pounding.

What happens next indicates where Carver locates the power to receive from a horrible contingency something good—in this case, the power to transform immense suffering into an experience of grace. The power is in the special capacity of persons who have experienced suffering to recognize others as persons, to sympathize with them, and then to offer those persons much needed hospitality. Though it is initially very brief, the couple's testimony of loss immediately fleshes them out to the baker. They are no longer customers but persons. Overcome with compassion at the weight of their loss, the baker immediately invites them to sit. He clears a space for them by shoving the adding machine and the telephone book aside (objects clearly symbolic of economies of exchange) and apologizing the best way he can. The baker confesses that his humanity has narrowed: "'Maybe once, maybe years ago, I was a different kind of human being. I've forgotten, I don't know for sure. But I'm not any longer, if I ever was. Now I'm just a baker'" (SGT, 87). Laboring for sustenance as a mere "jobholder" (Arendt's term) has reduced and alienated him from community; it has been difficult to be childless, lonely, and constantly working.[24] He appears to have no other contact with families except for the role he plays in their big private celebrations—birthdays and weddings—over and over again. "Just imagine all those candles burning" (SGT, 89). The baker plays a role in these celebrations but is never present for them. The celebrations may take place in his neighborhood, but the baker is not really a neighbor.

These are not minor details in this story. Borgmann begins his critique of technological society with a lament for the lack of significant public feasts and celebrations, events someone like the baker could be a part of, events with the power to turn people into neighbors. As the baker talks, the focus of the story shifts from the pain of the specific loss of the child toward the shared recognition of everyone's alienation from each other in contemporary life. There had been nothing between them, no cause for connection; the baker's two actions in the story to this point had been to bake and to call a customer in anger. But now his humanity is called for. He must become fully human to them if he is to meet them in their need.

And he responds. He participates in the giving and receiving of grace the best way that he knows how: by meeting the basic need they have for food.

"You probably need to eat something," the baker said. "I hope you'll eat some of my hot rolls. You have to eat and keep going. Eating is a small, good thing in a time like this," he said.

He served them warm cinnamon rolls just out of the oven, the icing still runny. He put butter on the table and knives to spread the butter. Then the baker sat down at the table with them. He waited. He waited until they each took a roll from the platter and began to eat. "It's good to eat something," he said, watching them. "There's more. Eat up. Eat all you want. There's all the rolls in the world in here." (SGT, 88)

They begin with cinnamon rolls but end with heavy, rich bread as their communion through talking and personal storytelling deepens. "'Smell this,' the baker said, breaking open a dark loaf. 'It's a heavy bread, but rich'" (SGT, 89). It is a full sensory experience: they smell and taste the bread; they see the light in the windows; they listen to each other. It is without doubt one of the most grace-filled scenes in contemporary American literature.

Despite the clear evocation of the Eucharist in this scene, some critics use this moment to suggest that Carver's grace here is purely human and "not Grace in the Christian sense at all."[25] But this interpretation both limits the profound depth of the scene and minimizes the Christian understanding of grace in all the ways it appears. The best way to avoid both mistakes might be to refer to Borgmann's three categories of grace and to illustrate how the scene evokes all three. The first is sacramental grace, that which specifically takes place in Christian worship. The second is actual grace, an example of which is the granting of forgiveness. The third is a concept that Borgmann learns from Karl Rahner: universal grace, which is the "omnipresent goodness of salvation that every human being is capable of realizing."[26] Borgmann extends this concept by renaming it habitual grace. Habitual grace is historically qualified, in that each epoch provides (or fails to provide) some kind of general "habitat" that enables grace.

The scene in which the baker breaks bread is certainly an example of actual grace, in that the baker specifically requests and receives forgiveness from Ann. But the way the other two forms of grace appear in this story requires more development. Regarding habitual grace, Borgmann borrows concepts from Heidegger to argue that "habitats" that enable grace are concealed and are beginning to disappear in the information age, particularly in affluent countries. The industrial revolution was successful in liberating Americans from most of the struggles against contingencies such as death from illness or the lack of proper nutrition or shelter, and this liberation must be seen as a good thing. But it also more nefariously shaped us, as technology moved from responding to necessity into the mere "procurement of pleasures and luxuries" by way of mechanization and commodification. In the age of necessity, we saw our needs being met as a kind of grace. Today we expect those needs to be met, and technology is employed entirely for pleasures of consumption.[27] Put another way, in a pre-industrial community, the baker's work was meeting an essential need, and the bread he baked was a grace to his neighbors. In a consumer culture, his products are optional pleasures at the whim of his customers' wants. He is more of a cog in the machine of consumer capitalism than a whole person whose gifts are greatly needed by others. Although the baker has not completely lost his trade to a factory, his performance of his tasks has become machine-like. He simply works all day to provide commodities for pleasure, symbolized in this story by "party food"—the birthday cake.

But the contingency disrupts this pattern, if only for a few days. Through the displacement provided by a devastating accident, wants become needs, and habitual grace becomes unconcealed. The three persons together enact what Borgmann calls a "focal practice." The word "focus" comes from the Latin word related to "hearth" and its orienting force in the pretechnological world.[28] Borgmann believes that recovery of the good life involves the establishment of focal practices of which the "culture of the table"—meals deliberately prepared and shared—is primary. The loaf the baker breaks at the end of the story is heavy and rich, with the taste of molasses and coarse grains. It has just come out of the baker's oven, not out of a factory oven, and the three people are seated together as they share it. Carver emphasizes that the baker served the family, and then he waited for them to eat. Nothing is bought or sold. Nothing is rushed. This scene does not represent a denial of technology but a foregrounding of something more basic. It is food removed from commodities for pleasure—the cake—and put in the realm of a life-giving gift—bread. In Borgmann's terms it is a meal as an enactment of generosity and gratitude, not of economic exchange.[29] Even more important, as they continue to eat, they continue to talk. "They talked on into the early morning, the high, pale cast of light in the windows, and they did not think of leaving" (SGT, 89). Their taking the time to share the bread together revalues the labor the baker had spent in preparing it. The baker is humanized, made into a whole person. The couple is also humanized by receiving a rich and deep hospitality at their moment of greatest need.

The giving and receiving of hospitality takes place here only because of the unconcealment of habitual grace. In other words, it is the very real suffering of the Weiss family that elicits a hospitable response from the baker. Responding to the other only out of obligation quickly reaches a limit in its ability to actually stir the subject to hospitable actions. But when one "suffers with" the other in sympathy, as Paul Ricoeur argues, a new equality is effected in the relationship between the parties in which each is truly able to see the other and themselves as irreplaceable—as an "each." That equality is ultimately based on the reminder of the vulnerability we have because of our mortality. "While equality is presupposed in friendship," Ricoeur writes, "in the case of the injunction coming from the other, equality is reestablished only through the recognition by the self of the superiority of the other's authority; in the case of sympathy that comes from the self and extends to the other, equality is reestablished only through the shared admission of fragility and, finally, of mortality."[30] The contingency, so obviously reminding all parties of their own mortality, provoked the shared admission of fragility that enabled a new relation to emerge between them.

Finally, Carver's evocation of sacramental grace is equally important in seeing this story as a response to the alienating patterns built into a technological society. Most Western readers are going to think about the Eucharist any time a character in a story deliberately breaks bread and shares it with others. There is enough of an echo of this practice in our culture to imbue the event with significance beyond itself. Although Carver's characters do not turn to God in this scene in any explicit way, there is recognition of one of the main purposes of the Eucharist. As commonly practiced in Christian worship, holy communion requires the participants to repair breaches in interpersonal relationships before partaking. It is meant to enact the sort of close family gathering that can occur only when grace in the form of real forgiveness has been offered and received by everyone present. Miroslav Volf argues that persons who celebrate the Eucharist reenact with each other God's forgiving embrace, through Christ, of those who offended him.[31] Richard Ford emphasizes that the commandment to have genuine communion with God and with others is the core of the Eucharist and that the event is "false if it is not connected with entering more fully into the contingencies and tragic potentialities of life in the face of evil and death. There can be no quick leap across Gethsemane and Calvary. Here are massive dislocation and disorientation, agonising loss and the demand to unlearn some of one's deepest convictions and habits."[32] Ann and the baker become agents of grace to each other, not just recipients of it. Volf explains that "inscribed on the very heart of God's grace is the rule that we can be its recipients only if we do not resist being made into its agents; what happens to us must be done by us. Having been embraced by God, we must make space for others in ourselves and invite them in— even our enemies."[33] Furthermore, as with the Eucharist, the grace is so rich in the scene because it was so costly a communion. In this case it took a violent contingency—the death of a child—for these people to look at each other in the face at all, let alone to take any responsibility for themselves or others.

To face another person is no small thing. Indeed, as I have already mentioned in this book, the notion of the face has been vital to current efforts within philosophy to recover a robust definition of justice.[34] Christianity should lead the way in this effort, because Christianity does not begin with a concept, but with a real face, a face that in the resurrected Christ was recognized in the breaking of the bread. Communities of the face, Ford reminds us, "can hardly be described in general terms" and Christianity has always been about seeing the glory of God in the particular face of Jesus Christ and in that image reflected in each person. Ford goes on to write that:

Christianity is characterised by the simplicity and complexity of facing: being faced by God, embodied in the face of Christ; turning to face Jesus Christ in faith; being members of a community of the face; seeing the face of God reflected in creation and especially in each human face, with all the faces in our heart related to the presence of the face of Christ; having an ethic of gentleness (*praütes*) towards each face; disclaiming any overview of others and being content with massive agnosticism about how God is dealing with them; and having a vision of transformation before the face of Christ 'from glory to glory' that is cosmic in scope, with endless surprises for both Christians and others.[35]

This "salvation, or health" is about "full hospitality and full worship," argues Ford. It is best fulfilled in feasting, the ultimate example of which is "to imagine an endless overflow of communication between those who love and enjoy each other."[36] It is also notably open to contingency as a way to experience personal transformation. To be sure, Carver's characters do not claim to see Christ in one another, and this story is no sermon. But their facing each other, breaking bread together, and listening to each other's stories in gentleness has turned them, finally, into neighbors.

The rich bread shared in fellowship and the recovery of contingency as a window to grace are not the only good things on offer here from Raymond Carver. The story itself is a small, good thing. The care in its crafting moves it from a commodity easily consumed for pleasure into a thing that requires a more involved response. "Thing" is not as bland of a word as it seems. When Carver describes his "second . . . post-drinking life," he says he developed a "belief in and love for the things of this world. Needless to say, I'm not talking about microwave ovens, jet planes, and expensive cars."[37] Heidegger pointed out that in Old High German, "thing" means a gathering, particularly a gathering to discuss matters of import.[38] For Heidegger, the work of art ideally provides that kind of orienting force, and it does so to an even greater extent if it focuses on the importance of simple things in our lives. So the interchangeability of the word "thing" in the title to mean bread, eating, communion, and the work of art serves to emphasize the need to return to these simple things that "flourish at the margins of public attention."[39] It is also to return to the notion of these simple things as gifts, which is to say, to return to them in their contingency.

It is not too much of a stretch to argue that the contingency of existence, the sheer gratuity of our life here together, is one of the main reasons why we have the arts at all. Artists are always asking, if only implicitly, why there is something rather than nothing. As Carver explains, "at the risk of appearing foolish, a writer sometimes needs to be able to just stand and gape at this or that thing—a sunset or an old shoe—in absolute and simple amazement."[40] A sunset or the nourishing simplicity of a shared loaf of bread does not particularly astonish us. So when the work of art brings us together to gape at it, it is a good thing indeed.

CHAPTER 8

The Lure of Transhumanism versus the Balm in *Gilead*

Marilynne Robinson's Redemptive Alternative

> *Love is the extremely difficult realization that something other than oneself is real.*
>
> —Iris Murdoch

Utopian fiction, when you can find the genuine article, is usually dreadful. Though aspects of Plato's *Republic* are considered utopian, the genre began with Thomas More's *Utopia* of 1516, which is interesting largely because it is not really utopian in the way we use the term today. Novels that earnestly represent a perfect society did not exist until the reform-minded nineteenth century, which produced Edward Bellamy's *Looking Backward* and William Morris's *News from Nowhere*, both socialist fantasies and both as dry as hay. Utopian novels took a new direction in the twentieth century, as in the behavioralist dream of B. F. Skinner's *Walden II* and the nebulous reification of "Good Being" that comprises Aldous Huxley's *Island*.[1] These books are dissatisfying primarily because the citizens populating them are barely a step away from cardboard cutouts, and none of the utopias provide blueprints that sane people would actually choose to pattern an ideal society around. These novelists seemed to have forgotten that Thomas More coined the word "utopia" sarcastically; it means "no place."[2]

The problem with all efforts to imagine a perfect world is that they usually require writers to imagine perfect people, and we have never known any. This partly explains why, although there are many visions of the ideal transhumanist future, none of them have been fleshed out in fiction.[3] Transhumanists are the twenty-first century's most vocal utopians, believing that humanity can and should move on to the next stage in evolution, including, eventually, the attainment of perfect health and immortality. Although transhumanism should not be equated with certain articulations of posthumanism,[4] it shares with them the view that human nature is plastic. In the words of Nick Bostrom, director of the Future of Humanity Institute and one of the most articulate proponents of transhumanism, humanity is "a work-in-progress, a half-baked beginning that we can learn to remold in desirable ways. Current humanity need not be the endpoint of evolution. Transhumanists hope that by responsible use of science, technology, and other rational means we shall eventually manage to become post-human, beings with vastly greater capacities than present human beings have."[5] But when it comes to illustrating why it would necessarily be better to have these greater capacities, Bostrom's descriptions remain abstract and theoretical; the closest thing he offers to an image of the future is his "Letter from Utopia" written by "your possible future self." The letter reads: "what you had in your best moment is not close to what I have now—a beckoning scintilla at most. If the distance between base and apex for you is eight kilometers, then to reach my dwellings would take a million light-year ascent. The altitude is outside moon and planets and all the stars your eyes can see. Beyond dreams. Beyond imagination. My consciousness is wide and deep, my life long."[6] The rest of the letter returns to explaining how to achieve that perfect self, which amounts to using technology to defeat mortality, suffering, and cognitive limitations.

Other proponents of transhumanism are equally abstract. The World Transhumanist Association, now calling itself Humanity+, recruits members to join the effort with its logo: "Healthier. Smarter.

Happier." By joining the organization, you will "show the world that there are lots of people who think that we can do better, that the current world is not the best that we can do, that there is room for improvement and that many human problems are solvable."[7] In other words, as the organization says on its website, "Humanity+ wants people to be better than well."[8] One of the principal members of Humanity+, a biogerontologist uncannily named Aubrey de Grey, has devoted his life to bio-regeneration research, the goal of which is to end physiological aging. Although admittedly there is no reason for medical science not to treat cellular aging like any other disease, or even to defeat aging as he defines it, the surrounding utopian rhetoric promises much more than these goals suggest. De Grey dedicates his book to "the tens of millions of people whose indefinite escape from aging depends upon our actions today" and refers to aging only as a disease, without any real engagement with substantive metaphysical objections to doing so.[9] In short, no one associated with transhumanism ever seems to question the main assumption that people who live longer with younger cells will necessarily live happier or more rewarding lives.

To the objection that the transhumanist vision remains unsatisfactorily abstract, Bostrom and others respond by arguing that this future world will be so different from what we know that we cannot comprehend or effectively illustrate it.[10] Because most things that transhumanists want (such as less suffering and healthier bodies) are goods in and of themselves, anyone who feels queasy about transhumanism soon discovers how difficult it is to question their vision. Indeed, fighting the transhumanists begins to feel like fighting a ghost—a perfectly happy, untouchable ghost. So the usual tactic is to argue that the transhumanist vision is undesirable because it is no longer recognizably human. This is what so-called bioconservatives like Leon Kass and Frances Fukuyama usually argue; namely, that there are certain fundamentals in human nature that should not be altered. Fukuyama's argument in *Our Posthuman Future* is exemplary of this approach; he argues that the biggest threat posed by biotechnology is that it will "alter human nature and thereby move us into a 'posthuman' stage of history." This is a threat for Fukuyama "because human nature exists, is a meaningful concept, and has provided a stable continuity to our experience as a species. It is, conjointly with religion, what defines our most basic values."[11]

The main problem with this argument is that it forgets that human nature has already been changed in countless ways by a variety of technologies, from vaccinating for illness to fighting depression with psychopharmaceuticals.[12] Where do you draw the line? Who gets to decide? But the weakest aspect of this approach is that it easily plays into the rhetorical strategies of typical transhumanist arguments. Most proponents of the unfettered development of enhancement technology lump all opposition to their ideas into a simple category of "people opposed to change."[13] Take this passage from the book *More Than Human: Embracing the Promise of Biological Enhancement,* by Ramez Naam. After a few short paragraphs about Fukuyama, Annas, and McKibben (each of whom argue for slowing down and regulating the advances of biotechnology), Naam writes:

> In short, a chorus of voices now argue that we should strive to preserve the status quo, that we should opt for stability over change, for the known over the unknown. To prevent unwanted changes, Fukuyama says, "We should use the power of the state" to restrict access to technologies that might undermine our current notions of humanity, that might allow individuals to surpass the mental and physical limitations we now know.
>
> This book is based on a very different premise: rather than fearing change, we ought to embrace it, rather than prohibiting the exploration of new technologies, society ought to focus on spreading the power to alter our own minds and bodies to as many people as possible. Rather than imposing a rigid view of what it means to be human on humanity, we ought to trust billions of individuals and families to decide that for themselves.[14]

This is a version of a nearly ubiquitous rhetorical strategy employed by proponents of unfettered biotechnology: you "biocons" all fear change; we "technoprogressives" embrace it. You want to control people via a rigid definition of humanity; we are in favor of individual liberties; and so on. This is an unfair characterization of, for example, Bill McKibben, who is as reform-minded as they come. But because McKibben worries about human nature changing for the worse, the argument he makes in his book *Enough* is treated as if he contends that we have had "enough" of technological developments that could see the end to cancer, or infant mortality, or any number of medical ills that we have faced. That is not his position on technology.[15]

In spite of the collection of straw men assembled and destroyed by technoprogressive thinkers, the most powerful critics of the transhumanist vision are not as fearful of change as they have been accused. Instead, these critics argue that the issue is not whether suffering is itself desirable or defensible; they all agree that suffering should be mitigated whenever possible. What the best critics argue instead is that transhumanism offers a false path to the good life. Transhumanism is wrong not because it promotes change, but because it promotes a dangerously thin definition of the good life, as if to be healthier, have a longer life, or experience less suffering will necessarily amount to a better life.[16] Furthermore, the danger is not in the actual future world that the transhumanists promise; the danger is in their present rhetoric. For what is at stake is not the "essence of humanity," but the present knowledge of what kind of good life it is actually possible and desirable for human beings, either as we are now or can ever become, to enjoy. While Naam and Bostrom and others want to "trust billions of individuals and families to decide for themselves" what the good life is, their best critics simply want individuals and families to know exactly what it is that they are choosing.

To see these choices as choices requires that we think them through to their possible ends. That kind of thinking requires narrative, which is the only place to flesh out what is both possible and desirable with regard to the "good life" for humans and posthumans alike. Margaret Atwood explains that this is the reason why we need utopian and dystopian fiction, for as a form, it "is a way of trying things out on paper first to see whether we might like them, should we ever have the chance to put them into actual practice."[17]

No one understands this fact better than Marilynne Robinson. In her Pulitzer Prize–winning novel *Gilead,* Robinson offers neither a dystopia nor a utopia, but a fully imagined alternative vision nonetheless. Her deeply Christian insistence is that the good life for both the individual and the community comes from learning first to see the beauty of our neighbors and then learning to love our neighbors as ourselves. For Robinson, striving for techno-utopia is not only unrealistic but also pushes the good life that is actually available to us further from view. The good life cannot be utopian because utopias depend on reforming society by fixing the people and institutions within it. The good life, instead, is the result of choosing to live by those virtues capable of transforming people into better lovers of others and thereby redeeming the community, one neighbor at a time.

A MATTER OF STYLE

As an experiment, one semester I taught *Gilead* at the end of a senior seminar centered around the question of what it means to be human in a posthuman world. I wanted to see what reading *Gilead* would feel like after students read *Frankenstein, Brave New World, Neuromancer, Snow Crash,* and many other works of speculative fiction. We read it slowly over the course of a few weeks as they worked on their final papers. When we talked about it, we were all astonished by how strong of an argument the style of *Gilead* makes; reading it felt like sitting still in a meadow after riding on a high-speed train. It is a kind of linguistic enactment of the central differences that Mark Helprin discusses in his illuminating essay "The Acceleration of Tranquility." In this essay he compares a 1906 man who conducts his overseas business by traveling by boat and writing letters to a 2016 man who does it all by high-speed jet and wireless devices. The 2016 man is more productive, but the 1906 man has time to ponder language, to be alone, and to simply be. Helprin argues that the 1906 man "was superior because he was allowed rest and reflection, his contemplation could seek its own level, and his tranquility was unaccelerated. While he was in his time a member of a privileged class unburdened by many practical necessities, today most Americans have similar resources and freedoms, and yet they, like their contemporaries in even the most exalted positions, have chosen a different standard."[18] Because it resists the acceleration of tranquility, *Gilead* is a countercultural novel. The novel requires the reader to slow down, to savor language, and to live inside of it. To read it is to believe that it is possible to choose to live in tranquility.

To stress this point, *Gilead* is an epistolary novel, a form chosen by some of the earliest pioneers of the novel but rarely by contemporary novelists. It consists of one letter written by an aged and dying father, John Ames, to his seven-year-old son. Slow, halting, and often quite repetitive, the letter is an intimate window into Ames's soul, into how he thinks about language, and into what the two have to do with each other. Ames often draws parallels between his life work as a pastor who writes down all of his sermons and the desire to pass some truths on to his son through this letter. For him writing is an intimate, life-giving, and even worshipful action. It is a spiritual dialogue. Ames writes to his son: "For me writing has always felt like praying, even when I wasn't writing prayers, as I was often enough. You feel that you are with someone. I feel I am with you now, whatever that can mean, considering that you're only a little fellow now and when you're a man you might find these letters of no interest."[19] While Emerson wrote that prayer is "the soliloquy of a beholding and jubilant soul," Ames believes it is necessarily a dialogue. Writing a letter, like praying, is an effort to connect with another person. It is an intimate conversation that is fraught with risks: Will the other understand? How will the other respond? There is already a kind of humility in this task of self-presentation and representation.

The epistolary form is, therefore, not a morally neutral choice. Martha Nussbaum argues that the question of style is central, not peripheral, to the writer's vision of the good life, as the "old quarrel between philosophy and literature is, as Plato clearly saw, not just a quarrel about ornamentation, but a quarrel about who we are and what we aspire to become." Style is itself a revelation, for to "choose a style is to tell a story about the soul."[20] And the story Robinson wants to tell is the slow and halting development of John Ames's soul, his gradual possession of the good life. The narrator is not David Copperfield, looking back on his childhood as an adult, but more like Francois Mauriac's Louis in *A Knot of Vipers,* who works through his feelings as he writes them in a series of death-bed letters. While Mauriac's story is a study in how a stubborn person who has resisted change his whole life can suddenly learn things through his own writing, Robinson's story traces the ongoing growth of a man who has always lived open to such change. Readers see John Ames wrestling with his impending death and mourning the time that he will not be able to share with his son. The reader also sees him imperfectly handle what is the central conflict of the novel: the sudden appearance of Jack (John Ames Boughton, the son of one of Ames's oldest friends) on the scene. Ames believes that Jack might be moving in on his young wife and son, in order to take his place when he dies. This situation provides Robinson with an opportunity for moral subtlety. Ames finds himself writing to his son as honestly as he might to a diary, but he is also trying to be restrained in his assessment of Jack in case Jack should end up being the child's father figure one day. Watching Ames walking that tightrope provides the novel with a surprising amount of what could be called spiritual drama.

It is because Ames takes his spiritual life, especially his responsibility to love others, so seriously that the letter moves from being merely a personal recounting of his life story to his son to being instruction for all readers in how to live the good life. The novel is an apology for the *vita contemplativa:* it reveals why the contemplative life is necessary; it illustrates the very real good that pursuing it accomplishes. John Ames has lived a good life—not perfect and not perfectly happy—but a *good* life. The difference between the perfect and the good is what matters.

THE VIRTUE OF HOPE

The epistolary form in Robinson's hands reveals that the good life lived in hope of personal change and growth diverges substantially from the consumer-oriented version of the good life as lived by the protean self. John Ames's views resemble those of Josef Pieper, who insisted that genuine hope is inexorably linked to the recognition of one's creaturely status. In *On Hope*, Pieper argues that the "innermost structure of created nature" is the state of being "on the way" or in the *status viatoris*. The virtue of hope is the "proper virtue of the 'not yet.'" The difference between a person being *status viatoris* and the transhumanist self remaking itself is that being *status viatoris* springs necessarily from having been created. Hope is a virtue dependent on recognizing this condition, for "in the virtue of hope more than in any other, man understands and affirms that he is a creature, that he has been created by God."[21] The virtue of hope comes from the acknowledgment that persons are on the way to some end as directed by someone else, not to constant change directed by the self. The completeness, or beatitude, that one is on the way to is fully human, not angelically transcendent. It will be started, but not completed, in this life. Ames reflects (*Gilead,* 115):

> There's a mystery in the thought of the re-creation of an old man as an old man, with all the defects and injuries of what is called long life faithfully preserved in him, and all their claims and all their tendencies honored, too, as in the steady progress of arthritis in my left knee. I have thought sometimes that the Lord must hold the whole of our lives in memory, so to speak. Of course He does. And "memory" is the wrong word, no doubt. But the finger I broke sliding into second base when I was twenty-two years old is crookeder than ever, and I can interpret that fact as an intimate attention, taking Herbert's view.

"Memory" is the wrong word for what Ames describes here because it implies that individual lives are merely remembered by God. Ames instead believes that God recollects and comprehends persons; God takes the whole of a human life, all experiences, good and bad, and makes sense of and honors them at the completion of one's life. It is a vision of change and development that does not view suffering or other challenges as roadblocks, but more as catalysts to the completion of a distinct person. Hope comes from the redemption of all of one's experiences, not the forgetting of them.

Even more important to Ames's notion of hope is the idea of a telos that is not derived solely from nature. Telos is the end destination, the completion of the project. Ames compares God's "memory" of the "whole of our lives" to the way the writer pays "intimate attention" to all of the details. It is a loving attention to how the parts make up the whole. For Robinson, the resulting completion, the whole, is not Hegelian, but it is aesthetic—and distinctly theological. It is theological first because it requires a seer who is an external creator, or "Wholly Other." Second, it is theological because it insists on a creator who is also a loving seer—a seer who calls what he sees "good." Indeed, naming creation "good" is what the God of Genesis did after he completed each stage of the work of creation. For Robinson, as for Flannery O'Connor, the parallel between the artist and God is located in this act of naming an other "good." Because the artist is not God, her creations participate in the larger creation by revealing the beauty of other beings. Similarly, in *Art and Answerability*, M. M. Bakhtin argues that seeing the way God sees (not creating the way he creates) is what art realizes.[22] The artist must see his character as "a beautiful *given*. . . . The author must see all of him in the fullness of the present and admire him as such."[23] The goodness comes from being and not from behavior, so that the characters are not perfect; they are loved.[24]

This idea speaks directly to Marilynne Robinson's prophetic vision for fiction and to how her novel can contribute to the conversation in bioethics. Her novel is an example of Walter Brueggemann's belief that current prophetic voices offer and nourish an alternative consciousness to the dominant consciousness. Robinson's theologically aesthetic view of hope, of the ultimate completion of the person with strengths and with faults, with health and with arthritis, is nearly opposite to the view that drives the biotechnological revolution forward. Most proponents of the unfettered development of enhancement technologies have either a vague vision of perfection (say, a transhumanist telos) or no interest at all in perfection (say, a posthumanist anti-telos). A transhumanist wants to take biological evolution into his own hands and move it toward a self-selected goal that includes the elimination of suffering and death.[25] A posthumanist wants change for its own sake, completely substituting late modern notions of process for modern notions of progress.[26] Neither vision interrogates or fully imagines the changes it wants technology to make possible beyond mere assertions that of course people will be happier when they are smarter, healthier, live longer, or have the freedom to morph their bodies into whatever new form they desire.[27]

For Marilynne Robinson, hope for human change looks radically different. It is not found in technological solutions to problems but in the old-fashioned word "sanctification": that John Ames would be transformed, slowly, haltingly, and imperfectly (in this life) into someone who looks more and more like Christ. Thus, Robinson insists that the first way toward the good life is to revitalize the recognition that persons must be on the way to something. As creatures, we humans are unable to effect that end ourselves, but we can choose to see it and to cooperate with it. To accept a telos is the only way to have hope.

To accept that there could be an appropriate telos to an individual life is also to accept the need for judgment to discern that telos. As Alisdair MacIntyre argues, the exercise of such judgment cannot be reduced to a "routinizable application of rules"; it requires the hard-won ability to discern between "what any particular individual at any particular time takes to be good for him and what is really good for him as a man."[28] Though MacIntyre and Robinson might disagree about the precise telos for which persons should aim, they would certainly agree that good judgment cannot come from nowhere; its development requires the cultivation of the virtues and of various contemplative practices. The need for this cultivation explains why *Gilead* is its own argument; to read it is to contemplate its countercultural suggestion of where the good life is actually to be found. In short, *Gilead* insists on contemplation. And as Hannah Arendt argues in *The Human Condition,* such contemplation is increasingly hard to come by because modern technological societies have substituted the *vita activa* (the active life) for the *vita contemplativa* (the contemplative life) characteristic of the ancients. Furthermore, through the reign of *techne,* modern humans have made a "matter-of-course identification of fabrication with action,"[29] as if "to do" means only to be "productive." As Arendt suggests, this fix-it attitude of the modern human, which focuses on the *how,* is at odds with the goals and methods of the *vita contemplativa,* which focuses on the *why.* Decisions that formerly had been directed by the contemplative life are now directed by a technological can-do attitude. For Arendt, what is lost is nothing less than the original motivation of science and the origin of philosophy: the wonder of Being. The contemplative life, argues Arendt, began with the "famous contention of Plato, quoted by Aristotle, that *thaumazein,* the shocked wonder at the miracle of Being, is the beginning of all philosophy."[30]

It is this lost "shocked wonder at the miracle of Being" that *Gilead* insists on recovering. Through its form and its content, *Gilead* teaches that the good life starts with wonder. John Ames learns that "wherever you turn your eyes the world can shine like transfiguration. You don't have to bring a thing to it except a little willingness to see. Only, who could have the courage to see it?" (*Gilead,* 245). John Ames does have this courage, but it was earned at a cost. The cost of the truly good life is sacrificial love.

THE ROLE OF SUFFERING IN LEARNING TO LOVE YOUR NEIGHBOR

So how did John Ames get to this point? And what does his own suffering have to do with it? *Gilead* insists that Ames's previous struggles, especially with loneliness, shaped him for the better—and for the worse. Ames's first wife died at a young age, after the death of their only child, and he did not remarry until he met Lila when he was much older. In the letter he reflects often on the long years of loneliness. While these years later gave him a strong sense of the miracle of his son's life, they also contributed to his struggle with "covetise." As a younger man, Ames was pained by being childless when his friend Boughton had so many. The bitter covetise was particularly strong when, while Ames was baptizing yet another child of his friend, Boughton gave his son the name "John Ames Boughton" (Jack). Ames resisted the gesture: "My heart froze in me and I thought, This is *not* my child—which I truly had never thought of any child before. I don't know exactly what covetise is, but in my experience it is not so much desiring someone else's virtue or happiness as rejecting it, taking offense at the beauty of it" (*Gilead,* 188). Because of this covetise, it took Ames a long time to forgive Boughton, and he immediately felt coldness toward his namesake. "I'll tell you a perfectly foolish thing," he writes, "I have thought from time to time that the child felt how coldly I went about his christening, how far my thoughts were from blessing him." What is more, Ames admits that he has "never been able to warm to him, never" (*Gilead*, 188).

Ames's relationship with Jack, the returning prodigal, is the central tension in the novel. Ames has always had a shepherd's concern for his flock, and he loves his congregation deeply. When his own son was born (seven years previous to the start of this novel), Ames had experienced a new gratitude that continues, now, to overwhelm him, giving him an even stronger sense of the beauty of the gift of life. But that strengthened sense of life as a gift clashes with old feelings of coldness toward Jack, a clash that is now brought to the forefront with Jack's reappearance. Ames, though he is aged and nearing death, has not finished growing in his ability to love others. Jack will be the ultimate test of Ames's faith and love. It is a test that he eventually passes, but not without a struggle.

That Ames learns to love the unlovable is how the novel offers such a powerful picture of an ethic of hospitality. Robinson draws on the biblical parable of the prodigal son in order to show how that parable can expand outside of filial relations to elevate love for the neighbor as the highest form of love. The parable has long been celebrated by Christian readers for its flexibility in addressing the issues of self-giving love.[31] The father, who loves both his obedient son and his prodigal son, lavishes grace when the prodigal returns and becomes both a picture of God's love for humanity and a model for how we are to love one another. It is flexible because any person could be seen to play each of the three roles: sometimes we are prodigal sons ourselves, sometimes we are fathers to our own prodigals, and sometimes we are the obedient older brother who must learn to abandon his resentment and love the prodigal brother anew. In *Gilead*, the most obvious prodigal is Jack Boughton, who leaves his own father, Old Man Boughton. He moves away from the community and pursues various relationships without concern for morality. But less obviously, John Ames writes about his own father and grandfather and their various relationships, and it becomes clear that they have each drifted from role to role themselves.

What makes *Gilead* unique in its reflection on the parable is that there are two families here. The primary plot is not of John Ames and

his son; in some ways that relationship is only a vehicle for the epistolary form. The primary plot concerns John Ames's relationship with
Jack Boughton: his *neighbor's* son. Thus Robinson reveals that the parable of the prodigal son is every bit as much about love for the neighbor
as it is love for the son. The ethical imperative of the parable is to see
and to love others as the heavenly Father does—that is, selflessly and ex-
pansively. Kierkegaard argued that one of Christianity's unique teach-
ings is that everyone is a neighbor and that "only when one loves the
neighbor, only then is the selfishness in preferential love rooted out and
the equality of the eternal preserved."[32] Robinson's illumination of the
prodigal son parable thus turns *Gilead* into a powerful alternative vision
to the dominant consciousness. As we have seen, the culture of enhancement promotes preferential love, not neighborly love. *Gilead* instead insists that only that love that learns how to see the "unlovable"
other as good and beautiful will lead to meaningful growth and change
for everyone involved.

John Ames seems to have long possessed a sense of wonder at the
beauty of existence, especially as it is found in the human face. But
the combined facts of the frailty of his own life and his struggle to see
the good in the unlovable Jack puts a finer point on it. The challenge is
what enables Ames to become a loving father. This passage reflects it
perfectly:

> Now, in my present situation, now that I am about to leave this
> world, I realize there is nothing more astonishing than a human
> face. Boughton and I have talked about that, too. It has something
> to do with incarnation. You feel your obligation to a child when
> you have seen it and held it. Any human face is a claim on you,
> because you can't help but understand the singularity of it, the
> courage and loneliness of it. But this is the truest of the face of an
> infant. I consider that to be one kind of vision, as mystical as any.
> (*Gilead*, 66)

In this passage, Ames seems to reason a lot like Emmanuel Levinas, in
that it is the face of the other that binds him to ethical action. But there
is a crucial difference here that speaks to the heart of the difference

between what fiction can offer to ethical reflection and what philosophy can offer. It is also a difference that speaks to what a Christian ethics of hospitality brings to the table that Levinasian ethics cannot. What happens to Ames when he sees Jack's face has "something to do with incarnation"; it is a mystical vision, very personal, and it comes when he faces his own mortality. There is nothing in Levinas that resembles this kind of vision, nothing that can convert the face of the other into an image of Christ that inculcates the virtue of love, not just the duty of obligation.[33] The Christian's astonishment at the beauty of a human face comes not just from the idea of the other's ethical demands but from the person of Christ working in John Ames to reveal the Christlike beauty of the other's face *and* the truth of the other's ethical demands.

Hans Urs von Balthasar develops the difference thoroughly in his book *Love Alone Is Credible*. He explains that "Christian action is above all a secondary reaction to the primary action of God toward man." This primary action, is of course, the incarnation and sacrificial death of Jesus Christ. Von Balthasar continues that "only by presupposing God's prior and inconceivable forgiveness can the limitations of human good will be transcended, and only thus can the danger of human pride be avoided: through God's love, I am first of all one who has been humbled, for my 'entire debt' had first to be forgiven, and my own secondary act of forgiveness is merely an echo. . . . This is not a principle of 'mere justice,' but the logic of absolute love." The difference between "absolute love" and "mere justice" is what separates Christian hospitality from the ethics of Levinas. When the Christian encounters the other, that other is a neighbor whom he has been commanded to love with the same love (and out of the same source of power) that he himself has received. What is more, the neighbor appears to the Christian as Christ himself. Von Balthasar explains that the "Christian encounters Christ *in* his neighbor, not beyond him or above him," and so love for the neighbor corresponds directly with the incarnation.[34]

While this kind of neighborly love is certainly the goal of Christian hospitality, that is not to say it is easy. Ames clearly struggles to see the beauty of Christ in his neighbor Jack Boughton. He prays continually and tries, above all else, to honor him. Like the older brother in the prodigal son parable, Ames struggles with envy; Jack's life is ongoing, but his own is coming to an end. But Ames's challenge is bigger than that of the prodigal's older brother because Jack is not returning in order to seek forgiveness for his dishonorable treatment of his family. Ames admits that he harbors a prejudice against those who "never really repent and never really reform" (*Gilead*, 156). Because honoring the other person is at the core of Ames's faith convictions, his struggle is intense: the real test of Ames's ability to honor another person comes literally in the face of one who is dishonorable. Ames reflects that the stakes are high and the potential rewards are great, "because at the root of real honor is always the sense of the sacredness of the person who is its object" (*Gilead*, 139).

John Ames wins the struggle. He comes out of himself to call Jack good and to see him as beautiful. Ames, broken himself and facing death, is able to separate the fact of Jack's inherent sanctity from his behavior, and blesses him. The scene is the novel's climax, but it is also beautifully underwritten in order to emphasize Ames's genuine humility. Jack is once again on his way out of town, leaving his father to die without him. But Jack has developed such a strong and new rapport with Ames (the proof of Ames's victory over his own inclinations to reject him) that Jack asks Ames to tell his father good-bye for him. With Jack's head resting almost against his hand, Ames blesses him with a benediction from the biblical book of Numbers: "The Lord make His face to shine upon thee and be gracious unto thee: The Lord lift up His countenance upon thee, and give thee peace." Ames adds: "Lord, bless John Ames Boughton, this beloved son and brother and husband and father" (*Gilead*, 241). That the Numbers benediction is perfectly expressive of Ames's feelings makes sense. In his own frailty, he has learned that although grace resides in the face of God, it can only be seen in the shining face of another person.

Although the blessing scene illustrates that Christian love for the neighbor has a validity that in no way depends on one's agreement to accept it or on the merit of the object, it also illustrates how the Christian's love for the neighbor rests on more than obligation or contract. It rests on grace. Having received this kind of love, individuals are thereby equipped and motivated to see the image of God in every face and to love each person represented thereby. Through John Ames, Robinson wrests love away from the language of contract and command and moves it toward the language of response to revelation and grace.[35] Ames reflects that "the terrible pleasure we find in a particular face can certainly instruct us in the nature of the very grandest love" (*Gilead*, 204). The novel shows that learning to love this way is a process not of one-time understanding of the demands of the face of the other, but of the character development of a person who repeatedly submits himself to trying to learn how to see others as God sees them: as good and inherently beautiful, but in need of redemption and healing. In the blessing, Ames lives out what he had reminded himself of before, when he was trying to reconcile himself to the possibility of Jack marrying his wife and taking his place: "Why do I worry so much over this Jack Boughton? Love is holy because it is like grace—the worthiness of its object is never really what matters" (*Gilead*, 209).

In comparing John Ames's vision to that of Levinas, I do not want to appear to misunderstand the difference between the goals of philosophy and the goals of theology or fiction. But when it comes to ethics, the limits of philosophy must be noted. Although a philosophical approach can insist on a "*direct* experience of the face and its claim,"[36] it can neither provide that experience nor adequately model it. As a theologically oriented novelist, Robinson seems to understand intuitively that it is one thing to argue that the face of the other makes an unconditional claim and quite another to learn to live it. *Gilead* is both understated and ambitious: it is only one man's private experience, but it serves as a powerful and attainable example of how a person can learn to love his neighbor as himself. The novel is a peek into how a character's moral imagination is formed and also a contribution to the formation of the moral imagination in its readers. This is why Robinson proclaimed in an interview that "art in a sense is occurring at the frontier of understanding because it integrates the problems of experience and the ordering of experience. Other conversations are farther from the essence of things; they should listen rather than talk so much."[37]

Part of what is at the "essence of things" for Robinson is that grace is not an abstract concept. Giving and receiving grace are experiences that are grounded in the reality of the holy Trinity. The relationship among the three men concerned—John Ames, Jack Boughton, and Old Man Boughton (Jack's father)—mirrors the Trinity, not in that the three participants are divine but in that they are in communion with one another as beings whose love gives the other the space to be different. Colin Gunton's point in stressing the Trinity as "three persons in communion, related but distinct" is to argue that when human life is viewed through Trinitarian relationships, what it means to be made in the image of God is seen differently, too. The human person does not find her being in particular qualities (such as consciousness) or in her separateness, but the human being is "one who is created to find his or her being in relation, first with other like persons but second, as a function of the first, with the rest of creation. . . . We find our reality in what we give to and receive from others in human community."[38]

To be a human created by God is to be created for loving relationship with God and other persons, not identity with God or other persons. The holy Trinity points the way to a dynamic loving relationship that is made stronger by the existence of more persons, more participants. It is meant to be an expanding circle. As Miroslav Volf puts it, "human beings come to be because Love, which is God, has 'projected' itself outside the rim of the Trinitarian circle so that there would be both objects and agents of love other than God. The love out of which human beings *come to be* as bearers of worth is fecund delight in the sheer 'thatness' of such creatures."[39] The idea that Jack can come to be more of what he was meant to be because of Ames's love is a central concern in *Gilead*. Ames's relationship with Jack neither overlooks the harm that Jack has caused, nor ceases, somehow, to hold him accountable. John Ames loves like the father in the prodigal son parable who gives freedom to the son, assuming (but not demanding) that the overwhelming grace will strengthen him so that he can change in his own way and time. Ultimately Ames's actions illustrate that only a broken person who knows that he has received grace and love has the humility to bless another person—and take responsibility for another person—toward whom he has never felt any natural affinity. Ames learns to love his neighbor as himself, Jack receives fatherly love, and the reader gets a picture of a man who has truly lived a good life.

THE RETURN TO THE HUMAN

Gilead is one of only three novels that Robinson has published in her over thirty-year career as a writer. In the space between the novels, she published one essay collection and a book called *Mother Country,* which exposes Britain's dumping of radioactive residue into the Irish Sea. The size and quality of this output suggests not only the obvious care Robinson takes with her prose but also her care in selecting subjects to tackle. Taken as a whole, her work shows consistent engagement with the question of the moral imagination: specifically, with how persons in Western culture are enculturated to interpret their experiences. This concern includes both a critique of the dominant paradigm that shapes the Western moral imagination and a persistently offered alternative to that paradigm.

Like Walker Percy, Robinson has narrowed in on the particularly dehumanizing dangers of a worldview that combines the certainty of scientific argument with the materialistic assumptions of scientific naturalism. *Absence of Mind,* one of her most recent books, addresses the subject directly. Her introduction pointedly blames the "self-declared rationalists" for creating an environment that denigrates humanity by the narrowness of its conception of life. Religion, she argues, "has always been the foil" for these thinkers, and it is "sometimes deplored as the sponsor of dysgenic compassion, sometimes as fomenter of oppression and violence."[40] Robinson does not defend religion per se. Instead, she continues the argument she began in her essay collection *The Death of Adam,* in which she points out the ethical ramifications of what she calls a "Malthus/Darwin/Nietzsche/Freud" paradigm. Such a paradigm is not morally neutral; it dictates an "ethic of competition and self-seeking." It doesn't engage religion except to describe and reject it.[41] Robinson is most concerned with the boldness that often accompanies the outright rejection of years of religious and theological thinking about what it means to be human. Ancient models are simply co-opted by a naturalistic model that (however unintentionally) tends to "lower us all in our own estimation."[42]

For Robinson, a view that debases people inevitably debases the moral imagination. She asks, "Where does this theory get its seemingly unlimited power over our *moral* imaginations, when it can rationalize stealing candy from babies—or, a more contemporary illustration, stealing medical care or schooling from babies—as readily as any bolder act? Why does it have the stature of science and the chic of iconoclasm and the vigor of novelty when it is, *pace* Nietzsche, only mythified, respectablized resentment, with a long, dark history behind it?"[43] Robinson's prophetic voice is strong here. She is not dismissing evolutionary science; she is urging readers to recognize the ramifications of a full-scale acceptance of the story that evolutionary science tells about who we are and why we are here.

As my reading of *Gilead* suggests, Robinson is not content simply to reject the hold that naturalistic evolutionary theory has on the moral imagination. She offers an alternative; she suggests a way back to the human from its prodigal, posthuman wanderings. The way to return to being human is not to win the scientific debate over origins, or the existence of God, or the biological essence of human nature. It is instead to return to "the cult of the soul, that old Jacob lamed and blessed in a long night of struggle."[44] To argue for a return to the "cult of the soul" is not a mere rhetorical point. Rather, it is to return to an ancient idea with the power to encourage humanity to be humane: that human existence, whatever else it may be found to be, is personal. And where persons are concerned, love is always the highest and best choice. Robinson is in accord with Hans Urs von Balthasar on this important point: "Only a philosophy of freedom and love can account for our existence, though not unless it also interprets the essence of finite being in terms of love. In terms of love—and not, in the final analysis, in terms of consciousness, or spirit, or knowledge, or power, or desire, or usefulness."[45] If the essence of finite being should be interpreted in terms of love, then the stories that we tell ourselves about ourselves are crucial. The stories that we tell ourselves about ourselves as human beings will determine our ethical, political, and personal decisions.

Can the stakes be any higher? For Robinson, our primary freedom as a species is the freedom to choose the stories that shape us. That

Robinson chooses to account for humanity in terms of a greater story of a loving creator God is what makes her a Christian and what motivates her as a novelist. Because whether they proceed from enlightenment confidence or postmodern insouciance, stories that describe human existence in terms of self-directed evolution are doomed to replace love with "respectablized resentment." But to tell a story that describes human existence in terms of love is to believe it is possible to live according to the demands of love, too.

I began this book with Ralph Waldo Emerson, the father of American transcendentalism, and Ray Kurzweil, his spiritual heir. Emerson believed that "a man is a god in ruins" and dedicated his career to exhorting the reversal of that fall. Kurzweil has dedicated his career to trying to find the technology to do it—literally. At the end of his documentary *Transcendent Man,* Kurzweil proclaims of our posthuman future that, "regardless what you call it, it will be the universe waking up. So does god exist? Well, I would say not yet." It never seemed to occur to Emerson, nor does it seem to occur to Kurzweil today, to question his instincts about whether or not this kind of search for divinity is ultimately good for human persons. It also never seemed to occur to either Emerson or Kurzweil that a human attempt at divinity might backfire. They have both ignored serious warnings against Promethean overreaching, such as Mary Shelley's *Frankenstein,* and humorous attempts to reveal humanity's bungling nature, such as Anthony Trollope's *Barchester Towers:* "Till we can become divine we must be content to be human, lest in our hurry for a change we sink to something lower." It also never seemed to occur to either Emerson or Kurzweil that all utopian visions contain a simple irony, an inherent logical flaw. Margaret Atwood explains that "man is by definition imperfect, say those who would perfect him. But those who would perfect him are themselves, by their own definition, imperfect. And imperfect beings cannot make perfect decisions."[46] Should we not, therefore, be cautious?

Though he will never make perfect decisions, Ray Kurzweil has predicted a lot of things correctly: the proliferation of the internet, the ubiquity of personal computers, the move to completely wireless devices, just to name a few. His most accurate prediction may have been when he wrote in 1999 that "the primary political and philosophical issue of the next century will be the definition of who we are."[47] But in spite of Kurzweil's tremendous intellect and prophetic powers, were he sitting with Emerson in a room with Robinson, Atwood, Shelley, Hawthorne, O'Connor, Percy, and Morrison, and were they debating the question of who we should become in the twenty-first century, I would side with the novelists. Though the worlds they come up with would certainly be odd—and in many ways, unrecognizable—they would at least be populated with real persons. And since I am already dreaming, I would like to conclude this book by imagining what Iris Murdoch, a novelist and a philosopher, might say if she walked into the above debate. I picture Emerson, brandy snifter in hand, looking out the window and proclaiming, "Let us treat the men and women as if they were real; perhaps they are." To which Murdoch would reply, unflinchingly, that love is the better way. "Love?" Emerson might reply. "The lover ascends to the highest beauty, to the love and knowledge of the Divinity!" "No, my dear Ralph," she would insist, "love is the extremely difficult realization that something other than oneself is real."[48]

Fiction is the art of love for persons. When it is done well, it attends so closely to the reality of persons that it makes their discovery possible. It is this work that we cannot do without. While new versions of humanity will certainly emerge, technological change will never circumvent the reality that society consists of persons who must learn to live with one another. Learning to love others, not to create a perfect future, is the hard work that lies before us.

Notes

1. Previous works include Raymond Kurzweil, *The Age of Intelligent Machines* (Cambridge, MA: MIT Press, 1990); and Raymond Kurzweil, *The Age of Spiritual Machines: When Computers Exceed Human Intelligence* (New York: Viking, 1999).

2. *Transcendent Man*, directed by Barry Ptolemy (Ptolemaic Productions, 2009).

3. Kurzweil also speaks about using technology to eventually bring his father, who died at age fifty-eight from heart disease, back to life.

4. Lee M. Silver, *Remaking Eden: Cloning and Beyond in a Brave New World* (New York: Avon Books, 1997), 17.

5. Thus my thesis opposes that of David Noble in *The Religion of Technology*, who argues that there is a fundamental continuity between the medieval conception of transcendence and the desire for transcendence through technology. As this book will demonstrate, the goals are categorically different precisely to the degree that the means are profoundly different. David F. Noble, *The Religion of Technology: The Divinity of Man and the Spirit of Invention* (New York: Knopf, 1997).

6. According to Marilynne Robinson, evolutionary theory thus argued produces "a conception of humanity that is itself very limited, excluding as it must virtually all observation and speculation on this subject that have been offered through the ages by those outside the closed circle that is called modern thought." Marilynne Robinson, *Absence of Mind: The Dispelling of Inwardness*

from the Modern Myth of the Self (New Haven, CT: Yale University Press, 2011), x.

7. This is why the mantra "Only evolve!" can make sense only when one fully accepts Darwin's assumptions about how the universe did evolve: that is, by random mutation and not intelligent design. See Allen Buchanen, "Why Evolution Isn't Good Enough," in *Better Than Human: The Promise and Perils of Enhancing Ourselves* (Oxford: Oxford University Press, 2011), 26–51.

8. As I will outline in chapter 5, Hannah Arendt described this change as the ascent of *homo faber*—human as maker—which replaced the ancient emphasis on the *vita contemplativa*: the contemplative life. Hannah Arendt, *The Human Condition* (Chicago: University of Chicago Press, 1998).

9. Eric Cohen, *In the Shadow of Progress: Being Human in the Age of Technology* (New York: Encounter Books, 2008), 21.

10. The family of Ted Williams, for example, reportedly had his head cryogenically preserved for future use. David Hancock, "Ted Williams Frozen in Two Pieces," *CBS News*, February 11, 2009, http://www.cbsnews.com /stories/2002/12/20/national/main533849.shtml.

11. Brent Waters, *From Human to Posthuman: Christian Theology and Technology in a Postmodern World*, Ashgate Science and Religion Series (Burlington, VT: Ashgate, 2006).

12. Surfdaddy Orca and R. U. Sirus, "Ray Kurzweil: The h+ Interview," *h+*, December 30, 2009, http://hplusmagazine.com/2009/12/30/ray-kurzweil -h-interview.

13. "Greenfield v. Kurzweil: The Great Debate," ITConversations, March 28, 2006, http://www.podfeed.net/episode/Greenfield+v.+Kurzweil+Biotech +Will+it+Save+Us+or+Hurt+Us/185841.

14. As I will discuss in chapter 1, John Brockman, the personality behind Edge.org, writes: "in 1975, [scientist Edward O.] Wilson . . . predicted that ethics would someday be taken out of the hands of philosophers and incorporated into the 'new synthesis' of evolutionary and biological thinking. He was right." "The New Science of Morality," Edge, July 20, 2010, http://edge.org.

15. Arendt, *The Human Condition*, 2–3. Bill McKibben put it more bluntly: "Understanding which chromosomes are responsible for the expression of which proteins doesn't give you any added insight into whether designer babies are a good idea, any more than figuring out how to make an atom bomb turns you into an expert on when or where you should drop it." Bill McKibben, *Enough: Staying Human in an Engineered Age* (New York: Times Books, 2003), 182.

16. A good place to begin is with Alvin Kernan, *What's Happened to the Humanities?* (Princeton, NJ: Princeton University Press, 1997).

17. Geoffrey Galt Harpham, "Beneath and Beyond the 'Crisis in the Humanities,'" *New Literary History: A Journal of Theory and Interpretation* 36, no. 1 (Winter 2005): 21–22.

18. Brian Stock, "Ethics and the Humanities: Some Lessons of Historical Experience," *New Literary History: A Journal of Theory and Interpretation* 36, no. 1 (Winter 2005): 11.

19. Ibid., 15.

20. Ibid.

21. Martha Craven Nussbaum, *Love's Knowledge: Essays on Philosophy and Literature* (New York: Oxford University Press, 1990), 21. See also Rita Charon, "The Ethical Dimensions of Literature: Henry James's *The Wings of the Dove*," in *Stories and Their Limits: Narrative Approaches to Bioethics*, ed. Hilde Lindemann Nelson (New York: Routledge, 1997), 91–112.

22. John Guillory, "The Ethical Practice of Modernity: The Example of Reading," in *The Turn to Ethics*, ed. Marjorie Garber, Beatrice Hanssen, and Rebecca L. Walkowitz (New York: Routledge, 2000), 34.

23. Guillory argues that "the political public sphere depends finally upon the cultivation of an intermediate ethical practice, upon the development of the capabilities of private citizens through their individual practices upon themselves. Reading can be such a practice, but only insofar as it is not reducible to a pure pleasure of consumption, or to the instrument of morality" (ibid., 44).

24. Steven Pinker, "The Stupidity of Dignity," *New Republic*, May 28, 2008, http://www.tnr.com/article/the-stupidity-dignity.

25. Two recent volumes describing this move include: Jane Adamson, Richard Freadman, and David Parker, *Renegotiating Ethics in Literature, Philosophy and Theory*, Literature, Culture, Theory, vol. 29 (Cambridge: Cambridge University Press, 1998); and Marjorie B. Garber, Beatrice Hanssen, and Rebecca L. Walkowitz, *The Turn to Ethics* (New York: Routledge, 2000). Parker has also written an important defense of the move in *Ethics, Theory, and the Novel* (Cambridge: Cambridge University Press, 1994).

26. Harpham, "Beneath and Beyond the 'Crisis,'" 35–36.

27. Ibid., 26.

28. Wayne C. Booth, *The Company We Keep: An Ethics of Fiction* (Berkeley: University of California Press, 1988), 38–39.

29. Nussbaum, *Love's Knowledge*, 259.

30. Thus my project is consistent with that of Eleonore Stump, whose recent book *Wandering in Darkness* argues that we must turn to narrative when it comes to questions that involve persons. Doing so also serves to remind us that "precise, compelling arguments are not everything. If we insist on rigor above everything else, we are in danger of getting it above everything else: a fossilized view of the world, unable to account for the richness of the reality in

which we live our lives." Eleonore Stump, *Wandering in Darkness: Narrative and the Problem of Suffering* (Oxford: Oxford University Press, 2010), 27.

31. It also explains why the humanities, though considered for some time to be in crisis in the academy, continue to fight for viability. As Harpham has recently argued, the belief that human history is a product of human intentionality and motivation is a core assumption in humanistic disciplines, though that assumption has been under considerable fire. The goal of the humanities is to understand the depths of human intentionality, which is "irreducibly complex; it can never be grasped by a single principle of explanation." Harpham, "Beneath and Beyond the 'Crisis,'" 29.

32. Ibid., 35.

33. Oscar Wilde made a number of derogatory quips about ethics, such as "the telling of beautiful untrue things, is the proper aim of art," to which Wayne Booth responds that "it takes no very deep reading to discover that Wilde's aim is to create a better kind of person—the kind who will look at the world and at art in a superior way and conduct life accordingly." Booth, *The Company We Keep*, 11.

34. Ben Marcus, "Ben Marcus Talks with George Saunders," in *The Believer Book of Writers Talking to Writers*, ed. Vendela Vida (San Francisco: Believer, 2005), 327. Some would even agree with John Updike's insistence that one of the novel's "habitual aims" is to "sharpen the reader's sense of vice and virtue" (quoted in Booth, *The Company We Keep*, 24).

35. Joseph Conrad, Preface to "The Nigger of the 'Narcissus,'" Project Gutenberg, http://www.gutenberg.org/files/17731/17731-h/17731-h.htm#2H_PREF.

36. As Martha Nussbaum puts it, novels "conduct a philosophical investigation into the good of a human being" (*Love's Knowledge*, 390). Eric Cohen points out that as soon as we ask the question, what do we live for?, we may be asking the central question of bioethics. Cohen, *In the Shadow of Progress*, 47.

37. Toby Warner, "Interview with George Saunders," *Boldtype*, 2008, http://boldtype.com/11995.html.

Introduction

1. Ralph Waldo Emerson, *Nature and Selected Essays,* ed. Larzer Ziff (New York: Penguin, 2003), 80–81.

2. Simon Young, *Designer Evolution: A Transhumanist Manifesto* (Amherst, NY: Prometheus Books, 2006), 15.

3. Emerson, *Nature and Selected Essays,* 360.

4. Young, *Designer Evolution,* 21–22.

5. Wendell Berry reminds readers that past a certain scale, "the person who makes a technological choice does not choose for himself alone, but for others; past a certain scale, he chooses for *all* others. . . . Past a certain scale, there is no dissent from a technological choice." Wendell Berry, *Standing by Words: Essays* (Washington, DC: Shoemaker & Hoard, 2005), 60.

6. Alasdair MacIntyre, *After Virtue: A Study in Moral Theory* (Notre Dame, IN: University of Notre Dame Press, 2007), 220. One of the most interesting new studies on the ethics of posthuman efforts to free the human self through technology is Ann Weinstone's *Avatar Bodies.* Weinstone's work reveals how posthuman ethics tend to retreat from the category of the person as person. She writes: "For those specifically working under the banner of posthumanism, hopes for mitigating human violence have expressed themselves, not only as productive engagements with difference and its vector, the nonhuman, but also as an aversion to speaking of human-to-human relationships at all. The cyborg is never a hybrid of two or more *people*" (*Avatar Bodies: A Tantra for Posthumanism,* Electronic Mediations, vol. 10 [Minneapolis: University of Minnesota Press, 2004], 6).

7. Discussing Levinas, Laurie Zoloth clarifies that the idea of autonomy itself is a good one, but when autonomy is "rendered only as dignity," it tends to forget its responsibility to others ("I Want You: Notes Toward a Theory of Hospitality," in *The Ethics of Bioethics: Mapping the Moral Landscape,* ed. Lisa A. Eckenwiler and Felicia Cohn [Baltimore: Johns Hopkins University Press, 2007], 207–8). Although I agree with this particular point, I will later argue in this book that Levinasian ethics has substantial limits in providing a foundation for true hospitality.

8. Jacques Maritain, *The Person and the Common Good* (New York: Charles Scribner's Sons, 1947), 47.

9. Jürgen Habermas, *The Future of Human Nature* (Malden, MA: Blackwell, 2003), 30.

10. Ibid., 52.

11. Ibid.

12. Ibid., 43.

13. Zoloth, "I Want You," 206–7.

14. Ibid., 207.

15. Miroslav Volf, *Exclusion and Embrace: A Theological Exploration of Identity, Otherness, and Reconciliation* (Nashville: Abingdon, 1996), 21.

16. I concur with Martha Nussbaum that to divide ethical approaches into "Kantian," "Utilitarian," and "Virtue Ethics" is a category mistake. My position in this book is closest to what she calls "anti-Utilitarian ethics," and agrees with three claims that constitute the common ground Nussbaum sets out. These claims are (1) that moral philosophy be concerned with the agent as

well as the action; (2) that it should "concern itself with motive and intention, emotion and desire"; and (3) that it should focus on "the whole course of the agent's moral life, its patterns of commitment, conduct, and also passion." See Martha C. Nussbaum, "Virtue Ethics: A Misleading Category?" *Journal of Ethics: An International Philosophical Review* 3, no. 3 (1999): 170.

17. MacIntyre, *After Virtue,* 220–21.

18. Young, *Designer Evolution,* 39.

19. Ibid., 246.

20. For example, Zygmunt Bauman, a Jewish sociologist, writes that "accepting the precept of loving one's neighbor is the birth-act of humanity. All other routines of human cohabitation, as well as the predesigned or retrospectively discovered norms and rules, are but a never-complete list of footnotes to that precept. We can go a step further and say that, since this precept is the preliminary condition of humanity, civilization, and civilized humanity, if this precept were to be ignored or thrown away, there would be no one extant to recompose that list or ponder its completeness" (*Does Ethics Have a Chance in a World of Consumers?* Institute for Human Sciences Vienna Lecture Series [Cambridge, MA: Harvard University Press, 2008], 32–33).

21. Thomas Aquinas, *Summa Theologica* (New York: Benziger Brothers, 1947), http://www.ccel.org/ccel/aquinas/summa.FP_Q20_A1.html.

22. George Parkin Grant, *Technology and Justice* (Notre Dame, IN: University of Notre Dame Press, 1986), 30.

23. Carl Elliott, "A New Way to Be Mad" *Atlantic Monthly* (December 2000), http://www.theatlantic.com/magazine/archive/2000/12/a-new-way-to-be-mad/4671/.

24. Albert Borgmann, *Technology and the Character of Contemporary Life: A Philosophical Inquiry* (Chicago: University of Chicago Press, 1984), 92.

25. I wrote this sentence before I read Neil Postman, who uses the same irresistible metaphor to make a similar point. See Neil Postman, *Technopoly: The Surrender of Culture to Technology* (New York: Vintage, 1993), 14.

26. Flannery O'Connor, *Collected Works* (New York: Viking, 1988), 268.

27. The definition of "prophet" I am relying on here is not explicitly religious but a general term for someone who advocates for ideas or theories in a visionary way.

28. Walter Brueggemann, *The Prophetic Imagination,* 2nd ed. (Minneapolis: Fortress, 2001), 3.

29. He then writes, "If I go further . . . I discover that in most of the operations of the mind, each American appeals to the individual exercise of his own understanding alone. . . . Nor is this surprising. The Americans do not read the works of Descartes, because their social condition deters them from speculative studies; but they follow his maxims because this very social

condition naturally disposes their understanding to adopt them" (*Democracy in America*, vol. 2, trans. Henry Reeve, Electronic Classics Series [Hazleton: Pennsylvania State University, 2002], 489, http://www2.hn.psu.edu/faculty/jmanis/toqueville/dem-in-america1.pdf).

30. N. Katherine Hayles, *How We Became Posthuman: Virtual Bodies in Cybernetics, Literature, and Informatics* (Chicago: University of Chicago Press, 1999), 2.

31. Ibid., 3.

32. Carl Elliott, *Better Than Well: American Medicine Meets the American Dream* (New York: Norton, 2003), 298–99.

33. Ibid., 299.

34. Ibid., 299–300.

35. Brent Waters, *From Human to Posthuman: Christian Theology and Technology in a Postmodern World* (Burlington, VT: Ashgate Pub, 2006), 98–99.

36. Grant, *Technology and Justice*, 11.

37. For an excellent introduction to the range of theoretical approach taken toward technology, see Elaine L. Graham, *Representations of the Post/Human: Monsters, Aliens and Others in Popular Culture* (Manchester, UK: Manchester University Press, 2002).

38. Francis Bacon, *The New Atlantis*, ed. Jim Manis, Electronic Classics Series (Hazleton: Pennsylvania State University, 2002), 31, http://www2.hn.psu.edu/faculty/jmanis/bacon/atlantis.pdf.

39. Eric Cohen, *In the Shadow of Progress: Being Human in the Age of Technology* (New York: Encounter Books, 2008), 15.

40. Borgmann, *Technology and the Character of Contemporary Life*, 41.

41. Lee M. Silver, *Challenging Nature: The Clash of Science and Spirituality at the New Frontiers of Life* (New York: Ecco, 2006); John Harris, *Enhancing Evolution: The Ethical Case for Making Better People* (Princeton, NJ: Princeton University Press, 2007); Ronald Michael Green, *Babies by Design: The Ethics of Genetic Choice* (New Haven, CT: Yale University Press, 2007); Gregory Stock, *Redesigning Humans: Our Inevitable Genetic Future* (Boston: Houghton Mifflin, 2002); Ramez Naam, *More Than Human: Embracing the Promise of Biological Enhancement* (New York: Broadway Books, 2005); Ray Kurzweil, *The Singularity Is Near: When Humans Transcend Biology* (New York: Viking, 2005).

42. Borgmann, *Technology and the Character of Contemporary Life*, 1.

43. Ibid., 61.

44. Elliott, *Better Than Well*, 298.

45. Brueggemann, *The Prophetic Imagination*, 27.

46. Bauman, *Does Ethics Have a Chance,* 53.

47. See, for example, Daniel Clement Dennett, *Darwin's Dangerous Idea: Evolution and the Meanings of Life* (New York: Simon & Schuster, 1995).

48. Alasdair MacIntyre, *Dependent Rational Animals: Why Human Beings Need the Virtues* (Chicago: Open Court, 1999).

49. Denise Levertov, *New & Selected Essays* (New York: New Directions, 1992), 148.

50. The parable is recorded in Luke 10:25–37. The quotations here are from the English Standard Version.

51. Indeed, "love is not concerned with qualities or nature or essences but with *persons*" (Maritain, *The Person,* 29; emphasis in original).

Chapter 1.   The Moral Imagination in Exile

1. Daniel Clement Dennett argues that to begin with the fact that humans are a product of evolution would provide a "naturalistic basis for sound ethical thinking" (*Darwin's Dangerous Idea: Evolution and the Meanings of Life* [New York: Simon & Schuster, 1995], 481).

2. John Brockman, *The Third Culture* (New York: Simon & Schuster, 1996), 17.

3. "The New Science of Morality Seminar," Edge, July 20, 2010, http://edge.org/event/seminars/the-new-science-of-morality.

4. Ibid.

5. Ibid.

6. Dennett, *Darwin's Dangerous Idea,* 21.

7. Ibid., 468 (emphasis in original).

8. "2008: What Have You Changed Your Mind About? Why?" Edge, January 10, 2008, http://edge.org/annual-question/what-have-you-changed-your-mind-about-why.

9. This preamble and the answers to the question are also collected in John Brockman, *What Have You Changed Your Mind About? Today's Leading Minds Rethink Everything* (New York: Harper Perennial, 2009), xxii.

10. These appear to be Brockman's views, and do not, of course, reflect the views of everyone the Edge group polls. For example, in his response to the question, Roger Highfield, editor of *New Scientist* magazine, answered: "I am a heretic. I have come to question the key assumption behind this survey: 'When facts change your mind, that's science.' This idea that science is an objective fact-driven pursuit is laudable, seductive and—alas—a mirage. Science is a never-ending dialogue between theorists and experimenters. But people are central to that dialogue. And people ignore facts. They distort them or select

the ones that suit their cause, depending on how they interpret their meaning. Or they don't ask the right questions to obtain the relevant facts. . . . Science is when, in the face of extreme scepticism, enough facts accrue to change lots of minds" ("2008").

11. Marilynne Robinson, *Absence of Mind: The Dispelling of Inwardness from the Modern Myth of the Self* (New Haven, CT: Yale University Press, 2011), xii, x, 129.

12. Ibid., 49–50.

13. According to John Hacker-Wright, "Murdoch challenges both the notions of the will and of the empirical world as found in contemporary naturalism, because they do not describe the conditions by which the world is knowable by us, and also because these notions artificially delimit the scope of our moral lives" ("Naturalism and the Good," in *Iris Murdoch and the Moral Imagination: Essays,* ed. M. F. Simone Roberts, Alison Scott-Baumann, and Luisa Muraro [Jefferson, NC: McFarland, 2010], 205).

14. Iris Murdoch and Peter J. Conradi, *Existentialists and Mystics: Writings on Philosophy and Literature* (New York: Penguin Books, 1999), 82.

15. Ibid.

16. Hacker-Wright, "Naturalism and the Good," 205.

17. George Lakoff and Mark Johnson, *Philosophy in the Flesh: The Embodied Mind and Its Challenge to Western Thought* (New York: Basic Books, 1999).

18. Mark Johnson, *Moral Imagination: Implications of Cognitive Science for Ethics* (Chicago: University of Chicago Press, 1993), ix–x.

19. Ibid., 2.

20. Lee M. Silver, *Challenging Nature: The Clash of Science and Spirituality at the New Frontiers of Life* (New York: Ecco, 2006), 147.

21. Ibid., 166, 170.

22. Ibid., 148.

23. Ibid., 149.

24. Flannery O'Connor, *Collected Works* (New York: Viking Press, 1988), 197–209. In the chapter parenthetical references to this source use the abbreviation *CW.*

25. W. E. Vine, *The Expanded Vine's Expository Dictionary of New Testament Words,* ed. John R. Kohlenberger with James A. Swanson, a special ed. (Minneapolis: Bethany House, 1984), 779.

26. Peter Singer, *Writings on an Ethical Life* (New York: Ecco, 2000), 129, 159.

27. Eric Cohen, *In the Shadow of Progress: Being Human in the Age of Technology* (New York: Encounter Books, 2008), 63.

28. Silver, *Challenging Nature,* 84, 87.

Chapter 2.   Aylmer's Moral Infancy

1. On the distinctions within enhancement technologies, see C. Ben Mitchell, "On Human Bioenhancements," Center for Bioethics and Human Dignity, Trinity International University, September 24, 2010, http://cbhd .org/content/human-bioenhancements.

2. Melanie Lefkowitz and Beth Whitehouse, "Teen Breast Enlargement Surgeries on the Rise: More Teens Having Breast-Augmentation Surgery, but Is It Safe?" *Roanoke Times,* April 9, 2008, http://www.roanoke.com/theedge /teenstories/wb/157464.

3. Susan Kriemer, "Teens Getting Breast Implants for Graduation," *Women's eNews,* June 6, 2004, http://www.womensenews.org/story/health /040606/teens-getting-breast-implants-graduation.

4. The President's Council on Bioethics, *Beyond Therapy: Biotechnology and Pursuit of Happiness* (New York: HarperCollins, 2003), 13.

5. Ibid., 15, 16.

6. Eric Cohen argues that "this question—what do we live for?—is perhaps the central question of bioethics. If biotechnology is driven by the desire to live *better* than we do now, then we need some idea of what a better life actually entails. We need some way to judge whether things have actually improved, especially if improvement means something more than simply living longer or being safer. And we need some idea of what human biology has to do with the good human life" (*In the Shadow of Progress: Being Human in the Age of Technology* [New York: Encounter Books, 2008], 47–48).

7. Charles Taylor, *Sources of the Self: The Making of the Modern Identity* (Cambridge, MA: Harvard University Press, 1989), 51.

8. Carl Elliott, *Better Than Well: American Medicine Meets the American Dream* (New York: Norton, 2003). Elliott argues that the desire of an American to find his or her "true self" has, paradoxically, fueled our unprecedented development and use of enhancement technology (xviii–xix).

9. Ibid., xx.

10. Leon R. Kass, ed., *Being Human: Core Readings in the Humanities* (New York: Norton, 2004), 3.

11. To view this tale as a cautionary tale is to insist that Hawthorne was tempted to become Aylmer but was committed not to be. Thus my interpretation of "The Birth-mark" is a bit at odds with that of Hawthorne's most recent biographer, Brenda Wineapple. Although Wineapple paints a picture of the newlyweds as poor but very happy together, she sees Hawthorne's early stories as suffused with a "deadly ambivalence about women and, more broadly, sexual bodies and fatherhood" (*Nathaniel Hawthorne: A Life* [New York: Knopf, 2003], 175).

12. Nathaniel Hawthorne and James McIntosh, *Nathaniel Hawthorne's Tales: Authoritative Texts, Backgrounds, Criticism* (New York: Norton, 1987), 118. In the chapter parenthetical references to this source use the abbreviation *NHT*.

13. John Lammers, "Powers' Eve Tempted: Sculpture and 'The Birth-Mark,'" *Publications of the Arkansas Philological Association* 21, no. 2 (1994): 41–58. Lammers explains that Hawthorne's earlier comparison of Georgiana to Hiram Powers's statue *Eve Tempted* indicates that Hawthorne considered Georgiana to be "actually already physically perfect, contrary to Aylmer's judgment" (46), so that Aylmer's later comparison of her to Pygmalion's statue indicates that Aylmer is a "weirdly sexually frustrated puritan male who already has a truly perfect and sexual woman—not a fallen Eve—yet is only gleeful about changing this woman, not about having sex with her" (56).

14. According to M. M. Bakhtin's theory of grotesque realism in literature, the body that has been rejected by the Platonic perfection of classical statuary is accepted fully in the grotesque. The former emphasizes perfection and stasis; the latter, imperfection and change. "The grotesque image reflects a phenomenon in transformation, an as yet unfinished metamorphosis, of death and birth, growth and becoming. The relation to time is one determining trait of the grotesque image. The other indispensable trait is ambivalence. For in this image we find both poles of transformation, the old and the new, the dying and the procreating, the beginning and the end of the metamorphosis" (*Rabelais and His World* [Bloomington: Indiana University Press, 1984], 24).

15. Quoted in Mary E. Rucker, "Science and Art in Hawthorne's 'The Birth-Mark,'" *Nineteenth-Century Literature* 41, no. 4 (1987): 446.

16. Rucker emphasizes that far from being manipulated, "Georgiana calmly and knowingly accepts her husband's point of view" (ibid., 452).

17. Martha Craven Nussbaum, *Love's Knowledge: Essays on Philosophy and Literature* (New York: Oxford University Press, 1990), 379, 381.

18. Another bit of evidence that Hawthorne himself was opposed to ever making his own wife, Sophia, feel this way can be seen in his early confiding in Margaret Fuller that, now that he is happily married, "he should be much more willing to die than two months ago, for he had had some real possession in life, but still he never wished to leave this earth. It was beautiful enough" (quoted in Wineapple, *Nathaniel Hawthorne,* 167).

19. Søren Kierkegaard, *Works of Love,* trans. David F. Swenson and Lillian Marvin Swenson (Princeton, NJ: Princeton University Press, 1946), 172.

20. Liz Rosenberg argues that "Aylmer never truly *sees* his wife; even when she is dying, he misperceives the true import of her symptoms." Rosenberg concludes that "The Birth-mark" stands alone as an example of a story written by Hawthorne as a newlywed who must learn to accept the imperfections of his

wife ("'The Best That Earth Could Offer': 'The Birth-Mark,' a Newlywed's Story," *Studies in Short Fiction* 30, no. 2 [Spring 1993]: 147).

21. Alistair I. McFadyen, *The Call to Personhood: A Christian Theory of the Individual in Social Relationships* (New York: Cambridge University Press, 1990). This pause is "an understanding connected with another's unique identity and social location (as a specific point of view and action). Instead of immediately opposing or denying the validity of an alternative understanding, an attempt is made to explore and comprehend it" (136).

22. "One does not care for this ground to make it a different place, or to make it perfect, but to make it inhabitable and to make it better. To flee from its realities is only to arrive at them unprepared." Wendell Berry, *Standing by Words: Essays* (Washington, DC: Shoemaker & Hoard, 2005), 92.

23. Berry explains that "when understood seriously enough, a form is a way of accepting and of living within the limits of creaturely life. We live only one life, and die only one death. A marriage cannot include everybody, because the reach of responsibility is short" (ibid., 93).

24. Ibid.

25. I am indebted to my colleague Tiffany Kriner for this argument.

26. Susan Bordo, *Unbearable Weight: Feminism, Western Culture, and the Body* (Berkeley: University of California Press, 1993), xxv.

27. As Eric Cohen argues, "our biology (our life) is both given and taken away. To experience this alienation is to look upon our bodies as if looking upon someone else or something else. This detachment is the basis for modern biotechnology" (*In the Shadow of Progress*, 57).

28. Susan Bordo's book *Unbearable Weight* is particularly valuable on this point. She argues that when it comes to interpreting the value of the female body, the most potentially fruitful way forward would be to avoid either "essentializing ontologies" or an unhealthy desire to transcend the body. "The most powerful revaluations of the female body have looked, not to nature or biology, but to the culturally inscribed and historically located body (or to historically developed *practices*) for imaginations of *alterity* rather than 'the truth' about the female body. . . . Without imaginations (or embodiments) of alterity, from what vantage point can we seek transformation of culture?" (41).

Chapter 3.   The Faces of Others

1. In the months following the contest, Boyle changed her image and sold millions of copies of her recording of "I Dreamed a Dream."

2. Although the terms "speculative fiction" and "science fiction" are under some debate because of the secondary status of science fiction, I find

Margaret Atwood's use of the terms most helpful. Atwood makes "a distinction between science fiction proper—for me, this label denotes books with things in them we can't yet do or begin to do, talking beings we can never meet, and places we can't go—and speculative fiction, which employs the means already more or less to hand, and takes place on Planet Earth" (*"The Handmaid's Tale* and *Oryx and Crake* in Context," *PMLA: Publications of the Modern Language Association of America* 119, no. 3 [2004]: 513).

3. Jacques Ellul, *The Technological Society,* trans. John Wilkinson (New York: Knopf, 1964), 223.

4. This commodification is perhaps most simply seen in the categorization "human resources."

5. Flannery O'Connor, *Collected Works* (New York: Viking Press, 1988), 805–6.

6. George Saunders, *In Persuasion Nation: Stories* (New York: Riverhead Books, 2006), 3. In the chapter parenthetical references to this source use the abbreviation *PN*.

7. Albert Borgmann, *Technology and the Character of Contemporary Life: A Philosophical Inquiry* (Chicago: University of Chicago Press, 1984), 105.

8. Corporations merely respond to the greater logic of technological society, which George Parkin Grant has made clear, that "the new technologies of both human and non-human nature have been the dominant responses to the crises caused by technology itself" (*Technology and Justice* [Notre Dame, IN: University of Notre Dame Press, 1986], 17).

9. Borgmann, *Technology and the Character of Contemporary Life,* 105.

10. Ellul, *The Technological Society,* 302.

11. Carl Elliott, *Better Than Well: American Medicine Meets the American Dream* (New York: Norton, 2003).

12. VTech, "VTech Debuts Innovative 2010 Products That 'Touch' the Future of Learning and Fun," February 11, 2010, http://www.vtechkids.com /assets/data/press/{FA989806-FB4C-496C-8F22-1A792265548D} /releases/2011-02-13-VTech_InnoPad_Release.pdf.

13. Borgmann, *Technology and the Character of Contemporary Life,* 53.

14. Amy Laura Hall, *Conceiving Parenthood: American Protestantism and the Spirit of Reproduction* (Grand Rapids, MI: Eerdmans, 2008), 16.

15. David Bahr, "George Saunders: Oppressing the Comfortable," *Publishers Weekly* 247, no. 33 (2000): 323.

16. Gilbert Meilaender, *Neither Beast nor God: The Dignity of the Human Person* (New York: Encounter Books, 2009), 48.

17. Emmanuel Levinas, *Totality and Infinity: An Essay on Exteriority,* trans. Alphonso Lingis, Duquesne Studies: Philosophical Series, vol. 24 (Pittsburgh: Duquesne University Press, 1969), 178.

18. Robert Spaemann, *Persons: The Difference between 'Someone' and 'Something'* (New York: Oxford University Press, 2006), 77.

19. Ronald Michael Green, *Babies by Design: The Ethics of Genetic Choice* (New Haven, CT: Yale University Press, 2007), 116.

20. Ibid., 134.

21. Bernadette Wegenstein, "Introduction," *Configurations: A Journal of Literature, Science, and Technology* 15, no. 1 (2007): 3.

22. Ibid., 1–2.

23. As Brenda R. Weber and Karen W. Tice point out ("Are You Finally Comfortable in Your Own Skin? The Raced and Classed Imperatives for Somatic/Spiritual Salvation in *The Swan*," *Genders* 49 [2009]: 10), *The Swan*

> continues as a media touchstone, since it is still broadcast, talked about, referenced, and parodied. The show is a highly successful export commodity (it was sold to more than 50 international media markets), redistribution (it airs in the U.S. on Fox Reality and the Style network), product (both boxed sets of the television and the spin-off book, *The Swan Curriculum,* are available for sale online and in stores), and social phenomenon (in addition to the reality celebrity that enabled participants to become models, television personalities, and cover girls, *The Swan's* experts and style gurus appear across the makeover canon, including on such shows as *10 Years Younger* and *How Do I Look?*).

24. See Michael Swanwick's introduction to James Tiptree Jr., *Her Smoke Rose Up Forever,* ed. Jeffrey D. Smith (San Francisco: Tachyon, 2004). Also, for a fascinating account of the disquiet that led Sheldon to hide her identity behind Tiptree for ten years, see Julie Phillips, *James Tiptree, Jr.: The Double Life of Alice B. Sheldon* (New York: St. Martin's, 2006).

25. Borgmann, *Technology and the Character of Contemporary Life,* 92.

26. Martin Heidegger, *The Question Concerning Technology, and Other Essays* (New York: Harper & Row, 1977), 303.

27. Kimberly Jackson notes that "the makers of *The Swan* have pushed the editing process to the point of parody. The content of each episode is so minimal, one wonders if there exist more than five minutes of actual footage from the narrative of each woman's journey through the program. Each contestant has about three lines, in which she outlines a major trauma, her reason for wanting to be a Swan, and some obstacle to her successful completion of the Swan program" ("Editing as Plastic Surgery: The Swan and the Violence of Image-Creation," *Configurations: A Journal of Literature, Science, and Technology* 15, no. 1 [Winter 2007]: 62).

28. Martha Craven Nussbaum, *Poetic Justice: The Literary Imagination and Public Life* (Boston: Beacon, 1995), xiii.

29. Wegenstein, "Introduction," 5.

30. Tiptree, *Her Smoke Rose Up Forever,* 43. In the chapter parenthetical references to this source use the short title "Plugged In."

31. René Girard, *Deceit, Desire, and the Novel: Self and Other in Literary Structure* (Baltimore: Johns Hopkins University Press, 1976).

32. Toni Morrison, *The Bluest Eye* (New York: Plume Book, 1994), 204.

33. Girard, *Deceit, Desire, and the Novel,* 43.

34. Jürgen Habermas, *The Future of Human Nature* (Malden, MA: Blackwell, 2003), 30.

35. This situation brings to mind the famous example of Greta Van Susteren, the sharp CNN reporter who initially resisted the culture of beauty but eventually gave in to getting a face lift.

36. Joss Whedon makes a similar point with a character (also named Paul) in the television series *Dollhouse.* See note 39 below.

37. Joel Garreau, *Radical Evolution: The Promise and Peril of Enhancing Our Minds, Our Bodies—and What It Means to Be Human* (New York: Doubleday, 2005).

38. Ray Kurzweil, *The Age of Spiritual Machines: When Computers Exceed Human Intelligence* (New York: Viking, 1999), 93.

39. Additionally, the main character who tries to help Caroline/Echo is an FBI agent named Paul. And like Paul in "The Girl Who Was Plugged In," he is shown to be trapped in mimetic desire, writing himself into the story as Caroline's savior. See episode 6, "The Man in the Street."

40. Geoffrey Galt Harpham, *On the Grotesque: Strategies of Contradiction in Art and Literature,* Critical Studies in the Humanities (Aurora, CO: Davies Group, 2006), 70.

41. Interview with Margaret Atwood, http://www.randomhouse.com /features/atwood/interview.html (no longer available).

42. Martha Craven Nussbaum, *Love's Knowledge: Essays on Philosophy and Literature* (New York: Oxford University Press, 1990).

43. Pamela Orosan-Weine, "The Swan: The Fantasy of Transformation Versus the Reality of Growth," *Configurations: A Journal of Literature, Science, and Technology* 15, no. 1 (Winter 2007): 31.

44. For an insightful reading on the goddess elements of *The Swan,* see Jackson, "Editing as Plastic Surgery," 55–76. "While the contestant is supposed to be turned into a goddess, a beauty queen to be worshipped by the culture that created her ideal, the structure of the show disallows this relation and thus empties the goddess of her power" (57).

45. As N. Katherine Hayles argues, "the dramatic climax makes clear that P. Burke has ceased to be an individual. Rather, she/it is a component in a cybersystem. . . . Any claim to possessive individualism based on ownership of

the self has thus been co-opted by a merger between corporate capitalism and communication technologies so potent that it operates in the intimate territory of nerve and muscle as well as global networks" (*My Mother Was a Computer: Digital Subjects and Literary Texts* [Chicago: University of Chicago Press, 2005], 85).

Chapter 4.   *The Scorned People of the Earth*

1. Gregory Stock, *Redesigning Humans: Our Inevitable Genetic Future* (Boston: Houghton Mifflin, 2002), 7.

2. Lee M. Silver, *Remaking Eden: Cloning and Beyond in a Brave New World* (New York: Avon Books, 1997), 17.

3. Stock, *Redesigning Humans,* 61.

4. Lauren Slater, "Dr. Daedalus," *The Best American Science Writing 2002,* ed. Matt Ridley (New York: HarperCollins, 2002), 5.

5. Ronald Michael Green, *Babies by Design: The Ethics of Genetic Choice* (New Haven, CT: Yale University Press, 2007), 11.

6. Toni Morrison, *The Bluest Eye* (New York: Plume Book, 1994), 209. In the chapter parenthetical references to this source use the abbreviation *TBE.*

7. Hannah Arendt, *The Human Condition* (Chicago: University of Chicago Press, 1998), 58.

8. Carl D. Malmgren argues that *The Bluest Eye* "is itself the text that counterpoints the missing primer lines. It makes 'Jane' visible and gives her a kind of being; it is the attempt of Claudia/Morrison to make the silence speak, to give voice to the voiceless" ("Texts, Primers, and Voices in Toni Morrison's *The Bluest Eye,*" *Critique: Studies in Contemporary Fiction* 41, no. 3 [Spring 2000]: 259).

9. Toni Morrison, *Conversations with Toni Morrison,* ed. Danille Taylor-Guthrie, Literary Conversations Series (Jackson: University Press of Mississippi, 1994), 103.

10. Green, *Babies by Design,* 53-80.

11. Morrison and Taylor-Guthrie, *Conversations,* 162.

12. Quoted in Stock, *Redesigning Humans,* 158.

13. Apparently, the idea of changing brown eyes to blue is no longer a sci-fi fantasy. According to the *Orlando Sentinel,* a California doctor claims he can, for about $5,000, use a laser to permanently remove the brown pigment from the iris and replace it with blue ("Doctor Says He Can Turn Brown Eyes Blue—Permanently," *Orlando Sentinel,* November 1, 2011, http://www.orlandosentinel.com/ktla-brown-eyes-blue,0,7886019.story).

14. Stephen Crane, *Prose and Poetry*, ed. J. C. Levenson, Library of America 18 (New York: Library of America, 1996), 1311.

15. Charles Rubin, "Human Dignity and the Future of Man," *Human Dignity and Bioethics: Essays Commissioned by the President's Council on Bioethics* (Washington, DC: President's Council on Bioethics, 2008), 164, 165.

16. Malmgren, "Texts, Primers, and Voices," 253.

17. Lynne Tirrell, "Storytelling and Moral Agency," *Journal of Aesthetics and Art Criticism* 48, no. 2 (Spring 1990): 117.

18. Trudier Harris, *Fiction and Folklore: The Novels of Toni Morrison* (Knoxville: University of Tennessee Press, 1991), 15.

19. Linda Dittmar writes that "this foregrounding of the unstable and constructed nature of knowledge, and of the collaborative processes which guide it, affirms the possibility of positive change. . . . Depicting and enacting ways we produce and re-produce ideology, the text reminds us that we can take charge of our future" ("'Will the Circle Be Unbroken?' The Politics of Form in *The Bluest Eye*," *Novel: A Forum on Fiction* 23, no. 2 [Winter 1990]: 142).

20. Tirrell, "Storytelling and Moral Agency," 124.

21. Thus Jane S. Bakerman argues that Claudia's story provides an illustration of her own stable family in contrast to that of Pecola's, who is clearly complicit in her self-destruction ("Failures of Love: Female Initiation in the Novels of Toni Morrison," *American Literature: A Journal of Literary History, Criticism, and Bibliography* 52, no. 4 [1981]: 543).

22. As Shelley Wong argues, Morrison's language has a physicality to it that serves to emphasize the connectedness of "all the manifestations of material being." Thus Morrison's prose, says Wong, "becomes a means of survival" ("Transgression as Poesis in *The Bluest Eye*," *Callaloo: A Journal of African American and African Arts and Letters* 13, no. 3 [Summer 1990]: 479).

23. Paul Ricoeur, *Oneself as Another* (Chicago: University of Chicago Press, 1992), 182, 183.

24. Gwendolyn Brooks, *Selected Poems* (New York: Harper & Row, 1963), 71.

25. Ricoeur, *Oneself as Another*, 192.

26. Richard Ford explains that Ricoeur's concept of conviction borrows universals from Kant and the particulars of the face from Levinas to create a different notion of the self and of self-esteem, that is, above all "recognition of responsibility [that] offers an interiority adequate to Levinas's exteriority and proximity of the face, and it culminates in a concept of conscience, the essence of which is attestation in conviction" (*Self and Salvation: Being Transformed* [Cambridge: Cambridge University Press, 1999], 93).

27. Ricoeur, *Oneself as Another*, 193; emphasis in original.

28. Green, *Babies by Design*, 71.

Chapter 5.   What Makes a Crake?

1. For a discussion of the development of twentieth-century dystopias, see Tom Moylan, *Scraps of the Untainted Sky: Science Fiction, Utopia, Dystopia* (Boulder: Westview Press, 2000).

2. This is why Francis Fukuyama, for example, draws repeatedly on *Brave New World* to illustrate that the greatest threats to democracy are still totalitarian; it is just that now the threat comes from social engineers who seize power through seducing people into their orderly society instead of forcing them. Francis Fukuyama, *Our Posthuman Future: Consequences of the Biotechnology Revolution* (New York: Farrar, Straus and Giroux, 2002), 5.

3. Ronald Bailey, *Liberation Biology: The Scientific and Moral Case for the Biotech Revolution* (Amherst, NY: Prometheus Books, 2005), 11–12; emphasis in original.

4. Ronald Michael Green, *Babies by Design: The Ethics of Genetic Choice* (New Haven, CT: Yale University Press, 2007), 2, 5.

5. Coral Ann Howells, "Margaret Atwood's Dystopian Visions: *The Handmaid's Tale* and *Oryx and Crake*," in *The Cambridge Companion to Margaret Atwood*, ed. Coral Ann Howells (Cambridge: Cambridge University Press, 2006), 163.

6. Human as maker, or *homo faber*, has been subsumed into *animal laborans*, his goals altered slightly, and thinking more about "toolmaking tools" than fabrications per se. This is why Danette DiMarco is correct to call Crake the "quintessential *homo faber*" and the "transformer of materiality into instruments" ("Paradise Lost, Paradise Regained: Homo Faber and the Makings of a New Beginning in Oryx and Crake," *Papers on Language and Literature: A Journal for Scholars and Critics of Language and Literature* 41, no. 2 [Spring 2005]: 170, 183).

7. DiMarco provides a thorough account of how the division of communities and labor match Arendt's explanation of late industrial society (ibid., 177).

8. Howells, "Margaret Atwood's Dystopian Visions," 164.

9. "The modern age's conviction that man can know only what he makes . . . brought forth the much older implications of violence inherent in all interpretations of the realm of human affairs as a sphere of making." Hannah Arendt, *The Human Condition* (Chicago: University of Chicago Press, 1998), 228.

10. Ibid., 309.

11. Ibid., 238.

12. Margaret Eleanor Atwood, *Oryx and Crake: A Novel* (New York: Random House, 2003), 100. In the chapter parenthetical references to this source use the abbreviation *OC*.

13. Any liberal arts PhD student can attest to a similar biting disparity between stipends available for humanities students verses stipends available for students in the sciences.

14. Jacques Ellul, *The Humiliation of the Word* (Grand Rapids, MI: Eerdmans, 1985), 127.

15. Wendell Berry, *Standing by Words: Essays* (Washington, DC: Shoemaker & Hoard, 2005), 52.

16. Technique, Jacques Ellul explains, is the "*totality of methods rationally arrived at and having absolute efficiency* (for a given stage of development) in *every* field of human activity" (*The Technological Society,* trans. John Wilkinson [New York: Knopf, 1964], xxv; emphasis in original).

17. Ellul, *The Humiliation of the Word,* 126–27.

18. Ibid., 213.

19. Toni Morrison, "The Bird in Our Hand: Is It Living or Dead?," Nobel Prize acceptance speech, Stockholm, Sweden, December 7, 1993, http://azer.com/aiweb/categories/magazine/63_folder/63_articles/63_morrison_nobel.html.

20. Margaret Eleanor Atwood, "Scientific Romancing: The Art of the Matter," Kesterton lecture, Carleton University, Ottawa, Ontario, January 24, 2004.

21. Ellul, *The Humiliation of the Word,* 115.

22. Crake is also particularly enamored of violent computer games like Barbarian Stomp and Extinctathon, games that, as J. B. Bouson points out, "turn mass destruction into an enjoyable spectacle" ("'It's Game Over Forever': Atwood's Satiric Vision of a Bioengineered Posthuman Future in *Oryx and Crake,*" *Journal of Commonwealth Literature* 39, no. 3 [2004]: 143).

23. Nicholas G. Carr, *The Shallows: What the Internet Is Doing to Our Brains* (New York: Norton, 2010), 91.

24. The lines she quotes (*OC,* 84) fit the novel perfectly:

Tomorrow, and tomorrow, and tomorrow.
Creeps in this petty pace from day to day,
To the last syllable of recorded time;
And all our yesterdays have lighted fools
The way to dusty death

25. For Emmanuel Levinas, the appearance of another person's face constitutes an ethical claim on those who see it. Levinas insists on the absolute alterity—the otherness—of this Other; it cannot be subsumed into the self's vision or possessed by the self, as can a mere image. In its separateness, the Other calls into question the self's freedom, which is essentially a call to justice. Justice cannot be done when the relationship to the other, with a full acknowl-

edgment of responsibility, goes unrecognized. "Society does not proceed from the contemplation of the true; truth is made possible by relation with the Other our master. Truth is thus bound up with the social relation, which is justice. Justice consists in recognizing the Other my master." (*Totality and Infinity: An Essay on Exteriority,* trans. Alphonso Lingis, Duquesne Studies: Philosophical Series, vol. 24 [Pittsburgh: Duquesne University Press, 1969], 72).

26. Levinas might say that internet videos encourage the moral irresponsibility of the mythical ring of Gyges, which renders the seer invisible to all he sees, ostensibly escaping responsibility (ibid., 173).

27. Ibid., 76.

28. The novel never clarifies whether the actual girl in the image was Oryx, but it remains a possibility. The important thing is that Jimmy believes it was her, and so has before him an embodiment of what the internet culture would have preferred to remain an image.

29. Sven Birkerts writes, "What Atwood's inventive treatment of first and last things lacks is a plausible psychological basis. The man who would play God, who would rewrite creation, needs to be something more than a knowingly enigmatic figure conjured onto the page" ("Present at the Re-Creation," review of *Oryx and Crake,* by Margaret Atwood, *New York Times,* May 18, 2003, http://www.nytimes.com/2003/05/18/books/present-at-the-re-creation .html?ref=bookreviews).

30. Paulina Borsook, *Cyberselfish: A Critical Romp through the Terribly Libertarian Culture of High Tech* (New York: PublicAffairs, 2000), 276.

31. Ibid., 15.

32. As Bouson points out, Crake has already shown an emotional void by being excited instead of horrified by his mother's death from a "trans-genetic staph," and by himself performing a trial run on his own stepfather ("'It's Game Over Forever,'" 146.

33. Christopher Lasch, *The Culture of Narcissism: American Life in an Age of Diminishing Expectations* (New York: Norton, 1991), 31–51.

34. Ibid., 243, 244.

35. See, for example, Daniel Clement Dennett, *Breaking the Spell: Religion as a Natural Phenomenon* (New York: Viking, 2006).

36. Richard Ford, *Self and Salvation: Being Transformed* (Cambridge: Cambridge University Press, 1999), 37.

37. See Paul Ricoeur, *Oneself as Another* (Chicago: University of Chicago Press, 1992), 140–68, and Robert Spaemann, *Persons: The Difference between 'Someone' and 'Something'* (Oxford: Oxford University Press, 2006), 148–63.

38. Hans Urs von Balthasar, *Love Alone Is Credible* (San Francisco: Ignatius, 2004), 142–43.

Chapter 6.   I Love Humanity, but I Don't Like You

1. Elizabeth Cohen, "CDC: Antidepressants Most Prescribed Drugs in U.S.," CNN, July 9, 2007, http://www.cnn.com/2007/HEALTH/07/09/anti depressants/index.html.

2. For example, see William J. Bailey, "FactLine on Non-Medical Use of Ritalin (Methylphenidate)," Indiana Prevention Resource Center at Indiana University, October 31, 1995, http://www.onelife.com/edu/indiana.html.

3. For example, Southerners getting "accent reduction" treatment; see Carl Elliott, *Better Than Well: American Medicine Meets the American Dream* (New York: Norton, 2003), 13.

4. Paul Elie records how, taken together, the writers challenged Percy's "scientific outlook." Paul Elie, *The Life You Save May Be Your Own: An American Pilgrimage* (New York: Farrar, Straus and Giroux, 2003), 138.

5. As Elliott writes, "when Percy looks at the unhappy man or woman in the suburbs, he sees something very different. He does not see a patient with a problem, but a person in a predicament. And part of that predicament is that the person has come to see herself in just the same way that the psychiatrist sees her: as a person in need of treatment. She has come to see herself as an organism whose well-being can be measured in terms of her mental health, her sexual happiness, the state of her body, the way she 'functions' at work and at home" (*Better Than Well,* 157).

6. Elliott writes "it is not a medical problem, but an existential problem. It is not the predicament of an organism in an environment, but that of a wayfarer who has lost her way" (ibid.).

7. Walker Percy and Patrick H. Samway, *Signposts in a Strange Land* (New York: Farrar, Straus and Giroux, 1991), 192.

8. Ibid., 272.

9. Neil Postman describes scientism as "the desperate hope, and wish, and ultimately the illusory belief that some standardized set of procedures called 'science' can provide us with an unimpeachable source of moral authority" (*Technopoly: The Surrender of Culture to Technology* [New York: Knopf, 1992], 162).

10. Percy and Samway, *Signposts in a Strange Land,* 151.

11. Jürgen Habermas, *The Future of Human Nature* (Malden, MA: Blackwell, 2003), 106–7.

12. The novelist, says Percy, feels "more and more like the canary being taken down the mine shaft with a bunch of hearty joking sense-making miners while he, the canary, is already getting a whiff of something noxious and is staggering around his cage trying to warn the miners, but they can't understand him nor he them" (Percy and Samway, *Signposts in a Strange Land,* 159).

13. Percy writes: "If the scientist's vocation is to clarify and simplify, it would seem that the novelist's aim is to muddy and complicate. . . . Since the novelist deals first and last with individuals and the scientist treats individuals only to discover their general properties, it is the novelist's responsibility to be chary of categories and rather to focus upon the mystery, the paradox, the *openness* of an individual human existence" (*The Message in the Bottle: How Queer Man Is, How Queer Language Is, and What One Has to Do with the Other* [1975; rpt., New York: Picador USA, 2000], 108).

14. Paul Ricoeur, *Oneself as Another* (Chicago: University of Chicago Press, 1992), 162.

15. As John Sykes points out, Percy leaves More at the end of *Love in the Ruins* having not forsaken his pride in science, which must have remained a temptation for Percy, too. *The Thanatos Syndrome* may have been his effort to work through that (*Flannery O'Connor, Walker Percy, and the Aesthetic of Revelation* [Columbia: University of Missouri Press, 2007], 135).

16. The first thing that Percy insists through the character of Bob Comeaux is that there are no plans for bettering humanity, no matter how appealing, that are not backed by a flawed human planner. Writing about the practice of selective contraception, or eugenics, C. S. Lewis reminds readers that this inevitably means that one generation of individuals is deciding for the next what it should value. "From this point of view," Lewis writes, "what we call Man's power over Nature turns out to be a power exercised by some men over other men with Nature as its instrument" (*The Abolition of Man; Or, Reflections on Education with Special Reference to the Teaching of English in the Upper Forms of Schools* [New York: Macmillan, 1967], 55).

17. Walker Percy, *The Thanatos Syndrome* (New York: Farrar, Straus and Giroux, 1987), 195. In the chapter parenthetical references to this source use the abbreviation *TS*.

18. Ronald Bailey, *Liberation Biology: The Scientific and Moral Case for the Biotech Revolution* (Amherst, NY: Prometheus Books, 2005), 12.

19. Ibid.

20. Stanley Hauerwas writes that it strikes him as odd that "in the name of eliminating suffering, we eliminate the sufferer" (*Suffering Presence: Theological Reflections on Medicine, the Mentally Handicapped, and the Church* [Notre Dame, IN: University of Notre Dame Press, 1986], 24).

21. Charles Rubin, "Human Dignity and the Future of Man," in *Human Dignity and Bioethics: Essays Commissioned by the President's Council on Bioethics* (Washington, DC: President's Council on Bioethics, 2008), 157.

22. Ibid., 165. Wendell Berry goes so far as to argue that it is impossible to be unselfish in planning someone else's future: "Love for the future is self-love—love for the present self, projected and magnified into the future, and it

is an irremediable loneliness" (*Standing by Words: Essays* [Washington, DC: Shoemaker & Hoard, 2005], 61).

23. For an insightful discussion of the misconceptions in Percy's theory of language, see Sykes, *Flannery O'Connor, Walker Percy, and the Aesthetic of Revelation,* 86–110.

24. The two characters are among good company in Catholic fiction: the so-called whiskey priests of novels like Graham Greene's *The Power and the Glory,* whose very weakness is intended to show that what matters when it comes to loving and ministering to others is not human execution but divine intervention and intention. See Mark Bosco, *Graham Greene's Catholic Imagination* (New York: Oxford University Press, 2005).

25. Percy and Samway, *Signposts in a Strange Land,* 375.

26. Ibid., 312.

27. Ibid., 312–13.

28. As I mentioned in the introduction, I am using the term "personalism" in a broad sense to encompass any Christian scholars interested in defining the person in a way that resists the dualism inherent in *cogito ergo sum* and resists the notion of the individual as defined by autonomy. This list includes Gabriel Marcel, Jacques Maritain, Emmanuel Mounier, and Etienne Gilson.

29. For a full treatment of this issue, see Christina Bieber Lake, *The Incarnational Art of Flannery O'Connor* (Macon, GA: Mercer University Press, 2005).

30. Søren Kierkegaard, *Works of Love,* trans. David F. Swenson and Lillian Marvin Swenson (Princeton, NJ: Princeton University Press, 1946), 271. Human love has a tendency to be rigid and domineering, but true love "loves every human being according to the person's distinctiveness. The *rigid, the domineering person* lacks flexibility, lacks the pliability to comprehend others; he demands his own from everyone, wants everyone to be transformed in his image, to be trimmed according to his pattern for human beings" (270).

31. Erasmo Leiva-Merikakis, *Love's Sacred Order: Four Meditations* (San Francisco: Ignatius, 2000), 131–32.

32. Percy, *The Message in the Bottle,* 281.

33. Sykes explains that sacramentalism is in inherent in his theory of language, "in the form of a metaphysical realism that posits a real connection between mind and reality by way of symbol" (*Flannery O'Connor, Walker Percy, and the Aesthetic of Revelation,* 108).

34. In his introduction to the collection *Persons, Divine and Human,* Christoph Schwöbel explains that all of the theologians therein start "from the personal particularity of persons as it is constituted in their personal relations in personal communion," not in order to illuminate more about God but to better understand personhood (Colin E. Gunton and Christoph Schwöbel,

Persons, Divine and Human: King's College Essays in Theological Anthropology [Edinburgh: T&T Clark, 1991], 12–13).

35. Jean Zizioulas, *Being as Communion: Studies in Personhood and the Church,* Contemporary Greek Theologians, vol. 4 (Crestwood, NY: St. Vladimir's Seminary Press, 1985), 18.

36. Robert Spaemann, *Persons: The Difference between 'Someone' and 'Something',* Oxford Studies in Theological Ethics (Oxford: Oxford University Press, 2006).

37. Spaemann, *Persons,* 77.

38. It also shows a trend away from More's earlier tendency toward the same kind of depersonalizing abstraction that Comeaux is especially guilty of. More must learn to sacrifice himself and become a listener, to participate in redemption-as-wholeness: "One must speak simply and humbly with others— or, rather, one must *listen* empathetically to what the other speaks, allowing one to know and feel with another as completely as possible, without conditions of acceptance" (Richard T. Martin, "Language Specificity as Pattern of Redemption in *The Thanatos Syndrome,*" *Renascence: Essays on Values in Literature* 48, no. 3 [Spring 1996]: 219).

39. Percy and Samway, *Signposts in a Strange Land,* 390.

40. In his book *On Hope,* Josef Pieper explains that the Christian virtue of hope is dependent on understanding that the human is *homo viator,* he is human on the way, and his "essence" is to be in the "process of becoming." To circumvent that is also to eliminate the condition necessary for hope ([San Francisco: Ignatius, 1986], 18).

41. More's own self discovery is as important as that of his patient here. As Lewis A. Lawson puts it, "Tom has learned not to think of himself as a genius—who thinks himself exempt from the change he imposes on others— but to use the genius he has been given to be Harry Stack Sullivan's 'participant-observer,' a flawed human being caring for another through language" ("Tom More: Walker Percy's Alienated Genius," *South Central Review* 10, no. 4 [Winter 1993]: 50).

Chapter 7.   Technology, Contingency, and Grace

This chapter originally appeared in *Christianity and Literature* 60, no. 2 (Winter 2011): 289–305, and is reprinted with permission.

1. C. S. Lewis, *The Abolition of Man; Or, Reflections on Education with Special Reference to the Teaching of English in the Upper Forms of Schools* (New York: Macmillan, 1967), 77.

2. Raymond Carver, *Cathedral: Stories* (New York: Vintage Books, 1983). In the chapter parenthetical references to this source use the abbreviation SGT.

3. Carver explained in an interview: "When I sit down to write, I literally start with a sentence or a line. I always have to have that first line in my head, whether it's a poem or a story. Later on everything else is subject to change, but that first line rarely changes" (Ewing Campbell, *Raymond Carver: A Study of the Short Fiction,* Twayne's Studies in Short Fiction Series, vol. 31 [New York: Macmillan International, 1992], 105).

4. Today, of course, the internet has made it possible to purchase everything one needs for existence without ever talking to a single person.

5. Albert Borgmann, *Power Failure: Christianity in the Culture of Technology* (Grand Rapids, MI: Brazos, 2003), 84. Borgmann explains that the paradigm rarely gets challenged in the quotidian realm because its dominant reason for existence is the promotion of comfort and security, and the pattern was "profoundly successful in establishing its norms as dominant and unquestionable" (85).

6. Ibid., 65.

7. In this chapter, I will focus on accidents (event contingencies) not as good in and of themselves but as reminders of the fact that we can never be in control of everything. But it should be noted that Borgmann is even more interested in the existence of contingency in the realm of science, which comes down to the question of why there should be something rather than nothing. He addresses this issue explicitly in the essay "Contingency and Grace in an Age of Science and Technology," *Theology Today* 59, no. 1 (2002): 6–20.

8. Ibid.

9. Individuals who live by the illusion of control can neither deal with accidents that puncture the illusion nor see good things as gifts for which to give thanks. To see nothing as a gift is an attitude shortly followed by denying the possibility of a giver. This is the outcome that concerns the writer of Proverbs 30, who prays, "give me neither poverty nor riches, but give me only my daily bread. Otherwise, I may have too much and disown you and say, 'Who is the LORD?'" (English Standard Version).

10. Martha Craven Nussbaum, *Love's Knowledge: Essays on Philosophy and Literature* (New York: Oxford University Press, 1990), 43, 44.

11. Günter Leypoldt argues that because of the "open" nature of Carver's writing, critics often simplify his use of epiphany in order to "demonstrate their respective aesthetic ends" ("Raymond Carver's 'Epiphanic Moments,'" *Style* 35, no. 3 [Fall 2001]: 531).

12. The Library of America's version of Carver's *Collected Stories* and Carol Sklenicka's biography, both published in 2009, shed considerable light on Lish's editing process.

13. Raymond Carver, *Collected Stories,* The Library of America, vol. 195 (New York: Library of America, 2009), 995. The restored story was published in Ploughshares in 1982 and later in Cathedral in 1983. Because Carver eventually conceded to Lish's changes and later downplayed the severity of them, a whole generation of Carver scholars assumed that the harsh minimalism of "The Bath" and other stories in *What We Talk About* was Carver's choice. Tess Gallagher, the poet with whom Carver was living at the time, has reported that Carver acquiesced because of Lish's "power of publication access," but that he always felt that the book "did not represent the main thrust of his writing, nor the true pulse and instinct in the work" (Carol Sklenicka, *Raymond Carver: A Writer's Life* [New York: Scribner, 2009], 359).

14. Carver, *Collected Stories,* 991. Carver was puzzled by people who liked "The Bath" better; he felt that it, like many of his early stories, contained "unfinished business" (Campbell, *Raymond Carver,* 101). He was also careful to remind readers that for a story to be "finished" does not mean that the writer has provided all the answers.

15. Sklenicka's biography emphasizes how Carver saw the day of his last drink as the beginning of a new life, a life he called "gravy" in a poem by that title (*Raymond Carver,* x). In a letter to his sister-in-law he wrote that "every day I have now is a real gift, and I look on it as a gift, another blessing" (365).

16. Borgmann, *Power Failure,* 78.

17. Gadi Taub argues that Carver believes "fellow feeling" through empathy to be the only way to bridge the inevitable distance between people. It is also the source of meaning in the stories: "Far from being personally subjective, meaning is dependent on the ability to see beyond, if not break, the boundaries of subjectivity. It requires, so to speak, a subjectivity of more than one subject, or, at any rate, a sense of sharing, however partially, someone else's feelings. The sense of meaning, in short, is grounded in empathy" ("On Small, Good Things: Raymond Carver's Modest Existentialism," *Raritan: A Quarterly Review* 22, no. 2 [Fall 2002]: 116).

18. Robert Spaemann, *Persons: The Difference between 'Someone' and 'Something'* (New York: Oxford University Press, 2006), 182.

19. Ibid., 183.

20. Jean-Luc Marion, *Prolegomena to Charity,* trans. Stephen E. Lewis, Perspectives in Continental Philosophy, vol. 24 (New York: Fordham University Press, 2002), 2.

21. Vigilante films rely on unfocused anger and desire for revenge that we have all experienced from having been hurt ourselves. The more entitled individuals feel to a world without contingency, the more these fantasies will appeal.

22. For example, the films *Changing Lanes* and *Crash* both illustrate how easy it is to lash out against a person who is just another face in the crowd.

23. Borgmann advises that "when cancer strikes or a car crashes, we should resist the uncomprehending anger that rises from the culture of transparency and control, and instead pray for the grace that allows us to accept what has come our way. I realize that we brush up against the problem of evil and theodicy here. . . . If I find consoling grace, the evil before me, my evil, no longer cries out for explanation or revenge. Such acceptance, however, is quite compatible with pressing for cancer research and greater highway safety" ("Contingency and Grace in an Age of Science and Technology," *Theology Today* 59, no. 1 [2002]: 19).

24. Hannah Arendt is chillingly relevant on the point: "For even now, laboring is too lofty, too ambitious a word for what we are doing, or think we are doing, in the world we have come to live in. The last stage of the laboring society, the society of jobholders, demands of its members a sheer automatic functioning, as though individual life had actually been submerged in the overall life process of the species and the only active decision still required of the individual were to let go, so to speak, to abandon his individuality, the still individually sensed pain and trouble of living, and acquiesce in a dazed, 'tranquilized,' functional type of behavior" (*The Human Condition* [Chicago: University of Chicago Press, 1998], 322).

25. Mark A. R. Facknitz goes on to say that "it is what grace becomes in a godless world—a deep and creative connection between humans that reveals to Carver's alienated and diminished creatures that there can be contact in a world they supposed was empty of sense or love" ("'The Calm,' 'A Small, Good Thing,' and 'Cathedral': Raymond Carver and the Rediscovery of Human Worth," *Studies in Short Fiction* 23, no. 3 [Summer 1986]: 296).

26. Borgmann, *Power Failure*, 74.

27. Borgmann explains: "Mechanization is the invention of some machinery that takes over the toils and burdens of providing some good, and the good, freed from its natural encumbrances, social burdens, and cultural ties, becomes available as a commodity for purchase and consumption" (*Contingency and Grace*, 76).

28. Albert Borgmann, *Technology and the Character of Contemporary Life: A Philosophical Inquiry* (Chicago: University of Chicago Press, 1984), 196. My use of the word "pretechnological" here parallels that of Borgmann and Heidegger. It is not to suggest that there was ever a time before technology, but only that the postindustrial world views itself through a technological framework. For the implications of how humanity is "always already" technological, see Elaine L. Graham, *Representations of the Post/Human: Monsters, Aliens and Others in Popular Culture,* Manchester Studies in Religion, Culture, and Gender (Manchester, UK: Manchester University Press, 2002).

29. Borgmann, *Technology and the Character of Contemporary Life*, 204.

30. Ibid., 192. Ricoeur explains how this relation is based more on weakness than on strength: "in true sympathy, the self, whose power of acting is at the start greater than that of the other, finds itself affected by all that the suffering other offers to it in return. For from the suffering other there comes a giving that is no longer drawn from the power of acting and existing but precisely from weakness itself. This is perhaps the supreme test of solicitude, when unequal power finds compensation in an authentic reciprocity of exchange, which, in the hour of agony, finds refuge in the shared whisper of voices or the feeble embrace of clasped hands" (191).

31. Miroslav Volf, *Exclusion and Embrace: A Theological Exploration of Identity, Otherness, and Reconciliation* (Nashville: Abingdon, 1996), 129.

32. Richard Ford, *Self and Salvation: Being Transformed* (Cambridge: Cambridge University Press, 1999), 147.

33. Volf, *Exclusion and Embrace,* 129.

34. For an example, see Emmanuel Levinas, *Totality and Infinity: An Essay on Exteriority*. trans. Alphonso Lingis, Duquesne Studies: Philosophical Series, vol. 24 (Pittsburgh: Duquesne University Press, 1969).

35. Ford, *Self and Salvation,* 24–25.

36. Ibid., 271.

37. Campbell, *Raymond Carver,* 100.

38. Martin Heidegger, *Poetry, Language, Thought* (New York: Harper & Row, 1975), 174.

39. Borgmann, *Technology and the Character of Contemporary Life,* 199.

40. Campbell, *Raymond Carver,* 94.

Chapter 8.   The Lure of Transhumanism versus the Balm in Gilead

1. My characterization of "earnest" utopian fiction distinguishes these idealistic texts from the "critical utopias" that flourished in the mid to late twentieth century. As Tom Moylan argues, these novels, including a number of feminist utopias such as *The Female Man,* by Joanna Russ, and *Woman on the Edge of Time,* by Marge Piercy, preserved the "subversive imaging of utopian society and the radical negativity of dystopian perception." Utopian writing was thereby "saved by its own destruction and transformation into the 'critical utopia'" (*Demand the Impossible: Science Fiction and the Utopian Imagination* [New York: Methuen, 1986], 10).

2. Margaret Atwood writes that "perhaps he meant to indicate that although his Utopia made more rational sense than the England of his day, it was unlikely to be found anywhere outside a book" (*Negotiating with the Dead: A Writer on Writing* [Cambridge: Cambridge University Press, 2002], 93).

3. The closest thing to it is the Culture series by Iain Banks and *Glasshouse* by Charles Stross, neither of which actually inhabit a future enhanced-to-perfect world and so cannot be called utopian. The Culture series imagines a "postscarcity" world run by artificial intelligence (AI). In it, the AI are chosen to be leaders precisely because they are "good"—defined, notably, as not "bad"; that is, they cannot be corrupted. The society has supposedly conquered suffering, death, and other ills. But when a society is perfect, nothing can really happen, so all the conflict comes from the Culture's interactions with the outside world, and the "perfect" world is usually only referenced, not visited. Stross's *Glasshouse* references a similar world; people born in 2050 can live forever by downloading their consciousness into new bodies. But to generate the conflict necessary for a novel, Stross sets all the action in a "glasshouse experiment" in which people from this future world are transported into a world that resembles the "dark ages." In other words, they are put in a world that looks like ours, so that things can begin to happen.

4. As Kristi Scott explains, transhumanism comes out of the Enlightenment humanistic project, whereas posthumanism represents "a shift away from humanism" ("Transhumanism vs./and Posthumanism," Institute for Ethics and Emerging Technologies, July 14, 2011, http://ieet.org/index.php/IEET/more /scott20110714).

5. Nick Bostrom, "Human Genetic Enhancements: A Transhumanist Perspective," *Journal of Value Inquiry* 37, no. 4 (2003): 493, http://www.nick bostrom.com/ethics/genetic.html.

6. Nick Bostrom, "Letter from Utopia," *Studies in Ethics, Law, and Technology* 2, no. 1 (2008): 1–7, http://www.nickbostrom.com/utopia.pdf.

7. Tom McCabe, "Humanity+ Aubrey de Grey Advocacy Prize," Humanity+, May 9, 2011.

8. Tom McCabe, "Humanity+ Parsons Conference Press Release," Humanity+, May 4, 2011.

9. Aubrey de Grey and Michael Rae, *Ending Aging: The Rejuvenation Breakthroughs That Could Reverse Human Aging in Our Lifetime* (New York: St. Martin's Griffin, 2007). For a response, see Gilbert Meilaender, "Thinking about Aging," *First Things: A Monthly Journal of Religion and Public Life,* April 1, 2011, http://www.firstthings.com/article/2011/03/thinking-about -aging.

10. For a response to this argument, see Charles Rubin, "Human Dignity and the Future of Man," in *Human Dignity and Bioethics: Essays Commissioned by the President's Council on Bioethics* (Washington, DC: President's Council on Bioethics, 2008), 158.

11. Francis Fukuyama, *Our Posthuman Future: Consequences of the Biotechnology Revolution* (New York: Farrar, Straus and Giroux, 2002), 7.

12. See John Harris, *Enhancing Evolution: The Ethical Case for Making Better People* (Princeton, NJ: Princeton University Press, 2007), and Elaine L. Graham, *Representations of the Post/Human: Monsters, Aliens and Others in Popular Culture* (Manchester, UK: Manchester University Press, 2002).

13. The biopolitics chart provided by the "technoprogressives" of the IEET (the Institute for Ethics and Emerging Technologies) is instructive; see "Overview of Biopolitics," http://ieet.org/index.php/IEET/biopolitics.

14. Ramez Naam, *More Than Human: Embracing the Promise of Biological Enhancement* (New York: Broadway Books, 2005), 5.

15. Bill McKibben, *Enough: Staying Human in an Engineered Age* (New York: Owl Books, 2003).

16. Even if happiness is accepted as the final arbiter of the good life, when transhumanist assumptions are unconcealed, they touch few of the things that current research is beginning to suggest actually do make people happier. Joshua Wolf Shenk reports on the research of George Vaillant, who completed one of the most comprehensive longitudinal studies available on the subject. The research revealed that happiness had more to do with relationships than health, longevity, or wealth ("What Makes Us Happy?" *Atlantic,* June 2009, http://www.theatlantic.com/magazine/archive/2009/06/what-makes-us-happy/7439/).

17. Atwood, *Negotiating with the Dead,* 95.

18. Mark Helprin, "Acceleration of Tranquility," in *Digital Barbarism: A Writer's Manifesto* (New York: Harper, 2009), 12.

19. Marilynne Robinson, *Gilead* (New York: Farrar, Straus and Giroux, 2004), 19. In the chapter parenthetical references to this source use *Gilead.*

20. Martha Craven Nussbaum, *Love's Knowledge: Essays on Philosophy and Literature* (New York: Oxford University Press, 1990), 259.

21. Josef Pieper, *On Hope* (San Francisco: Ignatius, 1986), 13, 21.

22. M. M. Bakhtin, *Art and Answerability: Early Philosophical Essays,* ed. Michael Holquist and Vadim Liapunov, trans. Vadim Liapunov and Kenneth Brostrom, University of Texas Press Slavic Series, vol. 9 (Austin: University of Texas Press, 1990). Although Bakhtin does not freely use the name "God" because of the restrictions he was under in Russia in the early 1920s when he was writing *Art and Answerability,* the implications are clear from his argument.

23. Ibid., 19; emphasis in original.

24. Hans Urs von Balthasar argues that in aesthetic seeing, what is important is givenness of the other being: "just as in mutual human love, where the other *as* other is encountered in a freedom that will never be brought under my control, so too in aesthetic perception it is impossible to reduce the appearing form [*Gestalt*] to my own power of imagination" (*Love Alone Is Credible* [San Francisco: Ignatius, 2004], 53).

25. The goal is somewhat paradoxically linked to evolution, since evolution itself follows a much more narrow range of goals linked to species survival, not the perfection of humanity per se.

26. I am indebted to Brent Waters for this point. Waters writes that "there are no given borders or boundaries that determine what it means to be human. That determination lies squarely and solely in what humans will themselves to be. The mastery of nature and human nature will only be complete, only be perfected, through human self-transformation. But transformed into what?" (*From Human to Posthuman: Christian Theology and Technology in a Postmodern World* [Burlington, VT: Ashgate, 2006], 49). One does not have to believe in an "essential" human nature to be troubled by the fact that this question—"transformed into what?"—is rarely raised in our culture today. And when a question is not intentionally raised, it is answered by market forces, typically without meaningful ethical restraints.

27. One of the baldest of such assertions can be found in an otherwise responsible book by John Harris. "Enhancements are so obviously good for us that it is odd that the idea of enhancement has caused, and still occasions, so much suspicion, fear, and outright hostility" (*Enhancing Evolution,* 36). To say that "enhancements are so obviously good for us" requires both a very narrow definition of enhancement and a particular vision of the good, neither of which are delineated by Harris.

28. Alasdair MacIntyre, *After Virtue: A Study in Moral Theory* (Notre Dame, IN: University of Notre Dame Press, 2007), 150.

29. Hannah Arendt, *The Human Condition* (Chicago: University of Chicago Press, 1998), 306.

30. Ibid., 302.

31. See especially Henri J. M. Nouwen, *The Return of the Prodigal Son: A Story of Homecoming* (New York: Image, 1994).

32. Søren Kierkegaard, *Works of Love,* trans. David F. Swenson and Lillian Marvin Swenson (Princeton, NJ: Princeton University Press, 1946), 44.

33. In an essay that considers the difference between Levinas and specifically theological descriptions of ethics, Merold Westphal explains that "Levinas is not invoking God as the one who commands us to love our neighbor. He is claiming that the face of the neighbor confronts us not as a contractual proposal to be negotiated but as an unconditional obligation. It is unconditional in that its validity depends in no way either upon our agreeing to accept it or in the Other's doing something to evoke or merit our compliance" ("Levinas and the Immediacy of the Face," *Faith and Philosophy: Journal of the Society of Christian Philosophers* 10, no. 4 [1993]: 494).

34. Von Balthasar, *Love Alone Is Credible,* 112–13, 114.

35. Michael Vander Weele argues that the novel's language "frees us, for a time, from the busyness of transaction, from the language of instrumentality.

Though not directly addressed to us, it seems to expand our conditions of being, or, less ambitiously, our sensibility" ("Marilynne Robinson's *Gilead* and the Difficult Gift of Human Exchange," *Christianity and Literature* 59, no. 2 [Winter 2010]: 220).

36. Westphal describes this as a goal of Levinasian ethics (*Levinas and the Immediacy of the Face,* 494; emphasis in original).

37. Thomas Schaub, "An Interview with Marilynne Robinson," *Contemporary Literature* 35, no. 2 (Summer 1994): 244.

38. Colin E. Gunton and Christoph Schwöbel, *Persons, Divine and Human: King's College Essays in Theological Anthropology* (Edinburgh: T&T Clark, 1991), 59.

39. Miroslav Volf, "God, Justice, and Love: The Grounds for Human Flourishing," *Books and Culture* (January/February 2009): 28; emphasis in original.

40. Marilynne Robinson, *Absence of Mind: The Dispelling of Inwardness from the Modern Myth of the Self* (New Haven, CT: Yale University Press, 2011), xii.

41. Furthermore, when theorists like Daniel Dennett argue for ethical implications for evolutionary theory, they are not doing science, but their work is, according to Marilynne Robinson, "sheltered under the immunities granted to science" so that it seems to be unassailable (*The Death of Adam: Essays on Modern Thought* [Boston: Houghton Mifflin, 1998], 37).

42. Robinson, *Absence of Mind,* 32.

43. Robinson, *The Death of Adam,* 67; emphasis in original.

44. Ibid.

45. Von Balthasar, *Love Alone Is Credible,* 143.

46. Margaret Atwood, *In Other Worlds: SF and the Human Imagination* (New York: Nan A. Talese/Doubleday, 2011), 140.

47. Ray Kurzweil, *The Age of Spiritual Machines: When Computers Exceed Human Intelligence* (New York: Viking 1999), 62.

48. While the interpersonal additions and emphases are mine, the quotes are taken directly from Emerson and Murdoch respectively; see Ralph Waldo Emerson, *Essays: First Series,* http://www.literaturepage.com/read/emerson essays1-91.html, and Iris Murdoch and Peter J. Conradi, *Existentialists and Mystics: Writings on Philosophy and Literature* (New York: Penguin Books, 1999). Murdoch continues: "Love, and so art and morals, is the discovery of reality" (215).

Bibliography

Adamson, Jane, Richard Freadman, and David Parker. *Renegotiating Ethics in Literature, Philosophy, and Theory*. Literature, Culture, Theory, vol. 29. Cambridge: Cambridge University Press, 1998.

Aquinas, Saint Thomas. *Summa Theologica*. New York: Benziger Brothers, 1947. http://www.ccel.org/ccel/aquinas/summa.FP_Q20_A1.html.

Arendt, Hannah. *The Human Condition*. 2nd ed. Chicago: University of Chicago Press, 1998.

Atwood, Margaret. "*The Handmaid's Tale* and *Oryx and Crake* in Context." *PMLA: Publications of the Modern Language Association of America* 119, no. 3 (2004): 513–17.

———. *In Other Worlds: SF and the Human Imagination*. New York: Nan A. Talese/Doubleday, 2011.

———. *Negotiating with the Dead: A Writer on Writing*. Cambridge: Cambridge University Press, 2002.

———. *Oryx and Crake: A Novel*. New York: Random House, 2003.

Bacon, Francis. *The New Atlantis*. Edited by Jim Manis. Electronic Classics Series. Hazleton: Pennsylvania State University, 2002. http://www2.hn.psu.edu/faculty/jmanis/bacon/atlantis.pdf.

Bahr, David. "George Saunders: Oppressing the Comfortable." *Publishers Weekly* 247, no. 33 (2000): 322–23.

Bailey, Ronald. *Liberation Biology: The Scientific and Moral Case for the Biotech Revolution*. Amherst, NY: Prometheus Books, 2005.

Bailey, William J. "FactLine on Non-Medical Use of Ritalin (Methylphenidate)." Indiana Prevention Resource Center at Indiana University, October 31, 1995. http://www.onelife.com/edu/indiana.html.

Bakerman, Jane S. "Failures of Love: Female Initiation in the Novels of Toni Morrison." *American Literature: A Journal of Literary History, Criticism, and Bibliography* 52, no. 4 (1981): 541–63.

Bakhtin, M. M. *Art and Answerability: Early Philosophical Essays.* Edited by Michael Holquist and Vadim Liapunov. Translated by Vadim Liapunov and Kenneth Brostrom. University of Texas Press Slavic Series, vol. 9. Austin: University of Texas Press, 1990.

———. *Rabelais and His World.* Translated by Hélène Iswolsky. Bloomington: Indiana University Press, 1984.

Bauman, Zygmunt. *Does Ethics Have a Chance in a World of Consumers?* Institute for Human Sciences Vienna Lecture Series. Cambridge, MA: Harvard University Press, 2008.

Berry, Wendell. *Standing by Words: Essays.* Washington, DC: Shoemaker & Hoard, 2005.

Birkerts, Sven. "Present at the Re-Creation," review of *Oryx and Crake,* by Margaret Atwood. *New York Times,* May 18, 2003. http://www.nytimes.com/2003/05/18/books/present-at-the-re-creation.html?ref=bookreviews.

Booth, Wayne C. *The Company We Keep: An Ethics of Fiction.* Berkeley: University of California Press, 1988.

Bordo, Susan. *Unbearable Weight: Feminism, Western Culture, and the Body.* Berkeley: University of California Press, 1993.

Borgmann, Albert. "Contingency and Grace in an Age of Science and Technology." *Theology Today* 59, no. 1 (2002): 6–20.

———. *Power Failure: Christianity in the Culture of Technology.* Grand Rapids, MI: Brazos, 2003.

———. *Technology and the Character of Contemporary Life: A Philosophical Inquiry.* Chicago: University of Chicago Press, 1984.

Borsook, Paulina. *Cyberselfish: A Critical Romp through the Terribly Libertarian Culture of High Tech.* New York: PublicAffairs, 2000.

Bosco, Mark. *Graham Greene's Catholic Imagination.* American Academy of Religion Academy Series. Oxford: Oxford University Press, 2005.

Bostrom, Nick. "Human Genetic Enhancements: A Transhumanist Perspective." *Journal of Value Inquiry* 37, no. 4 (2003): 493–506. http://www.nickbostrom.com/ethics/genetic.html.

Bouson, J. B. "'It's Game Over Forever': Atwood's Satiric Vision of a Bioengineered Posthuman Future in *Oryx and Crake.*" *Journal of Commonwealth Literature* 39, no. 3 (2004): 139–56.

Brockman, John. *The Third Culture.* New York: Simon & Schuster, 1995.

———. *What Have You Changed Your Mind About? Today's Leading Minds Rethink Everything.* New York: Harper Perennial, 2009.

Brooks, Gwendolyn. *Selected Poems.* New York: Harper & Row, 1963.

Brueggemann, Walter. *The Prophetic Imagination.* 2nd ed. Minneapolis: For-
    tress, 2001.

Buchanen, Allen. *Better Than Human: The Promise and Perils of Enhancing
    Ourselves.* Oxford: Oxford University Press, 2011.

Campbell, Ewing. *Raymond Carver: A Study of the Short Fiction.* Twayne's Stud-
    ies in Short Fiction Series, vol. 31. New York: Twayne, 1992.

Carr, Nicholas G. *The Shallows: What the Internet Is Doing to Our Brains.* New
    York: Norton, 2010.

Carver, Raymond. *Cathedral: Stories.* Vintage Contemporaries. New York: Vin-
    tage Books, 1983.

———. *Collected Stories.* The Library of America, vol. 195. New York: Library
    of America, 2009.

Charon, Rita. "The Ethical Dimensions of Literature: Henry James's *The
    Wings of the Dove.*" In *Stories and Their Limits: Narrative Approaches to Bio-
    ethics,* edited by Hilde Lindemann Nelson, 91–112. New York: Routledge,
    1997.

Cohen, Elizabeth. "CDC: Antidepressants Most Prescribed Drugs in U.S."
    CNN, July 9, 2007. http://www.cnn.com/2007/HEALTH/07/09/anti
    depressants/index.html.

Cohen, Eric. *In the Shadow of Progress: Being Human in the Age of Technology.*
    New York: Encounter Books, 2008.

Conrad, Joseph. Preface to "The Nigger of the 'Narcissus.'" Project Gutenberg.
    http://www.gutenberg.org/files/17731/17731-h/17731-h.htm#2H
    _PREF.

Crane, Stephen. *Prose and Poetry.* Edited by J. C. Levenson. Library of America
    18. New York: Library of America, 1996.

de Grey, Aubrey, with Michael Rae. *Ending Aging: The Rejuvenation Break-
    throughs That Could Reverse Human Aging in Our Lifetime.* New York: St.
    Martin's, 2007.

Dennett, Daniel Clement. *Breaking the Spell: Religion as a Natural Phenomenon.*
    New York: Viking, 2006.

———. *Darwin's Dangerous Idea: Evolution and the Meanings of Life.* New
    York: Simon & Schuster, 1995.

DiMarco, Danette. "Paradice Lost, Paradise Regained: *Homo Faber* and the
    Makings of a New Beginning in *Oryx and Crake.*" *Papers on Language and
    Literature: A Journal for Scholars and Critics of Language and Literature* 41,
    no. 2 (Spring 2005): 170–95.

Dittmar, Linda. "'Will the Circle Be Unbroken?' The Politics of Form in *The
    Bluest Eye.*" *Novel: A Forum on Fiction* 23, no. 2 (Winter, 1990): 137–55.

Elie, Paul. *The Life You Save May Be Your Own: An American Pilgrimage.* New
    York: Farrar, Straus and Giroux, 2003.

Elliott, Carl. *Better Than Well: American Medicine Meets the American Dream.* New York: Norton, 2003.

——. "A New Way to be Mad." *Atlantic Monthly* (December 2000). http://www.theatlantic.com/magazine/archive/2000/12/a-new-way-to-be-mad/4671/.

Ellul, Jacques. *The Humiliation of the Word.* Translated by Joyce Main Hanks. Grand Rapids, MI: Eerdmans, 1985.

——. *The Technological Society.* Translated by John Wilkinson. New York: Knopf, 1964.

Emerson, Ralph Waldo. *Nature and Selected Essays.* Edited by Larzer Ziff. New York: Penguin, 2003.

Facknitz, Mark A. R. "'The Calm,' 'A Small, Good Thing,' and 'Cathedral': Raymond Carver and the Rediscovery of Human Worth." *Studies in Short Fiction* 23, no. 3 (Summer 1986): 287–96.

Ford, Richard. *Self and Salvation: Being Transformed.* Cambridge Studies in Christian Doctrine. Cambridge: Cambridge University Press, 1999.

Fukuyama, Francis. *Our Posthuman Future: Consequences of the Biotechnology Revolution.* New York: Farrar, Straus and Giroux, 2002.

Garber, Marjorie B., Beatrice Hanssen, and Rebecca L. Walkowitz. *The Turn to Ethics.* Culture Work. New York: Routledge, 2000.

Garreau, Joel. *Radical Evolution: The Promise and Peril of Enhancing Our Minds, Our Bodies—and What It Means to Be Human.* New York: Doubleday, 2005.

Gibson, William. *Neuromancer.* New York: Ace, 2004.

Girard, René. *Deceit, Desire, and the Novel: Self and Other in Literary Structure.* Translated by Yvonne Freccero. Johns Hopkins paperback ed. Baltimore: Johns Hopkins University Press, 1976.

Graham, Elaine L. *Representations of the Post/Human: Monsters, Aliens and Others in Popular Culture.* Manchester Studies in Religion, Culture, and Gender. Manchester, UK: Manchester University Press, 2002.

Grant, George Parkin. *Technology and Justice.* Notre Dame, IN: University of Notre Dame Press, 1986.

Green, Ronald Michael. *Babies by Design: The Ethics of Genetic Choice.* New Haven, CT: Yale University Press, 2007.

"Greenfield v. Kurzweil: The Great Debate." ITConversations, March 28, 2006. http://www.podfeed.net/episode/Greenfield+v.+Kurzweil+Biotech+Will+it+Save+Us+or+Hurt+Us/185841.

Guillory, John. "The Ethical Practice of Modernity: The Example of Reading." In *The Turn to Ethics,* edited by Marjorie Garber, Beatrice Hanssen, and Rebecca L. Walkowitz, 29–46. New York: Routledge, 2000.

Gunton, Colin E., and Christoph Schwöbel. *Persons, Divine and Human: King's College Essays in Theological Anthropology.* Edinburgh: T&T Clark, 1991.

Habermas, Jürgen. *The Future of Human Nature*. Malden, MA: Blackwell, 2003.

Hacker-Wright, John. "Naturalism and the Good." In *Iris Murdoch and the Moral Imagination: Essays,* edited by M. F. Simone Roberts and Alison Scott-Baumann, 203–20. Jefferson, NC: McFarland, 2010.

Hall, Amy Laura. *Conceiving Parenthood: American Protestantism and the Spirit of Reproduction*. Grand Rapids, MI: Eerdmans, 2008.

Hancock, David. "Ted Williams Frozen in Two Pieces." *CBS News,* February 11, 2009. http://www.cbsnews.com/stories/2002/12/20/national/main533849.shtml.

Harpham, Geoffrey Galt. "Beneath and Beyond the 'Crisis in the Humanities.'" *New Literary History: A Journal of Theory and Interpretation* 36, no. 1 (Winter 2005): 21–36.

———. *On the Grotesque: Strategies of Contradiction in Art and Literature*. Critical Studies in the Humanities. Aurora, CO: Davies Group, 2006.

Harris, John. *Enhancing Evolution: The Ethical Case for Making Better People*. Princeton, NJ: Princeton University Press, 2007.

Harris, Trudier. *Fiction and Folklore: The Novels of Toni Morrison*. Knoxville: University of Tennessee Press, 1991.

Hauerwas, Stanley. *Suffering Presence: Theological Reflections on Medicine, the Mentally Handicapped, and the Church*. Notre Dame, IN: University of Notre Dame Press, 1986.

Hawthorne, Nathaniel. *Nathaniel Hawthorne's Tales: Authoritative Texts, Backgrounds, Criticism*. Selected and edited by James McIntosh. A Norton Critical Edition. New York: Norton, 1987.

Hayles, N. Katherine. *How We Became Posthuman: Virtual Bodies in Cybernetics, Literature, and Informatics*. Chicago: University of Chicago Press, 1999.

———. *My Mother Was a Computer: Digital Subjects and Literary Texts*. Chicago: University of Chicago Press, 2005.

Heidegger, Martin. *Poetry, Language, Thought*. Translated by Albert Hofstadter. His Works. New York: Harper & Row, 1975.

———. *The Question Concerning Technology, and Other Essays*. Harper Colophon Books. New York: Harper & Row, 1977.

Helprin, Mark. "Acceleration of Tranquility." In *Digital Barbarism: A Writer's Manifesto,* 1–20. New York: Harper, 2009.

Howells, Coral Ann. "Margaret Atwood's Dystopian Visions: *The Handmaid's Tale* and *Oryx and Crake*." In *The Cambridge Companion to Margaret Atwood,* edited by Coral Ann Howells, 161–75. Cambridge: Cambridge University Press, 2006.

Huxley, Aldous. *Brave New World*. New York: Harper Perennial, 2006.

Institute for Ethics and Emerging Technologies (IEET). "Overview of Biopolitics." http://ieet.org/index.php/IEET/biopolitics.

Jackson, Kimberly. "Editing as Plastic Surgery: *The Swan* and the Violence of Image-Creation." In "Reality Made Over," ed. Bernadette Wegenstein, special issue, *Configurations: A Journal of Literature, Science, and Technology* 15, no. 1 (Fall 2007): 55–76.

Johnson, Mark. *Moral Imagination: Implications of Cognitive Science for Ethics.* Chicago: University of Chicago Press, 1993.

Kass, Leon R., ed. *Being Human: Core Readings in the Humanities.* New York: Norton, 2004.

Kernan, Alvin, ed. *What's Happened to the Humanities?* Princeton, NJ: Princeton University Press, 1997.

Kierkegaard, Søren. *Works of Love.* Translated by David F. Swenson and Lillian Marvin Swenson. Princeton, NJ: Princeton University Press, 1946.

Kriemer, Susan. "Teens Getting Breast Implants for Graduation." *Women's eNews* June 6, 2004. http://www.womensenews.org/story/health/040606/teens-getting-breast-implants-graduation.

KTLA News. "Doctor Says He Can Turn Brown Eyes Blue—Permanently." *Orlando Sentinel,* November 1, 2011. http://www.orlandosentinel.com/ktla-brown-eyes-blue,0,7886019.story.

Kurzweil, Ray(mond). *The Age of Intelligent Machines.* Cambridge, MA: MIT Press, 1990.

———. *The Age of Spiritual Machines: When Computers Exceed Human Intelligence.* New York: Viking, 1999.

———. *The Singularity Is Near: When Humans Transcend Biology.* New York: Viking, 2005.

Lake, Christina Bieber. *The Incarnational Art of Flannery O'Connor.* Macon, GA: Mercer University Press, 2005.

Lakoff, George, and Mark Johnson. *Philosophy in the Flesh: The Embodied Mind and Its Challenge to Western Thought.* New York: Basic Books, 1999.

Lammers, John. "Powers' Eve Tempted: Sculpture and 'The Birth-Mark.'" *Publications of the Arkansas Philological Association* 21, no. 2 (1994): 41–58.

Lasch, Christopher. *The Culture of Narcissism: American Life in an Age of Diminishing Expectations.* New York: Norton, 1991.

Lawson, Lewis A. "Tom More: Walker Percy's Alienated Genius." *South Central Review* 10, no. 4 (Winter 1993): 34–54.

Lefkowitz, Melanie, and Beth Whitehouse. "Teen Breast Enlargement Surgeries on the Rise: More Teens Having Breast-Augmentation Surgery, but Is It Safe?" *Roanoke Times,* April 9, 2008. http://www.roanoke.com/theedge/teenstories/wb/157464.

Leiva-Merikakis, Erasmo. *Love's Sacred Order: Four Meditations.* San Francisco: Ignatius, 2000.

Levertov, Denise. *New & Selected Essays.* New York: New Directions, 1992.

Levinas, Emmanuel. *Totality and Infinity: An Essay on Exteriority.* Translated by Alphonso Lingis. Duquesne Studies: Philosophical Series, vol. 24. Pittsburgh: Duquesne University Press, 1969.

Lewis, C. S. *The Abolition of Man; Or, Reflections on Education with Special Reference to the Teaching of English in the Upper Forms of Schools.* New York: Macmillan, 1967.

Leypoldt, Günter. "Raymond Carver's 'Epiphanic Moments.'" *Style* 35, no. 3 (Fall 2001): 531–47.

MacIntyre, Alasdair. *After Virtue: A Study in Moral Theory.* Notre Dame, IN: University of Notre Dame Press, 2007.

———. *Dependent Rational Animals: Why Human Beings Need the Virtues.* Chicago: Open Court, 1999.

Malmgren, Carl D. "Texts, Primers, and Voices in Toni Morrison's *The Bluest Eye.*" *Critique: Studies in Contemporary Fiction* 41, no. 3 (Spring 2000): 251–62.

Marcus, Ben. "Ben Marcus Talks with George Saunders." In *The Believer Book of Writers Talking to Writers,* edited by Vendela Vida, 313–32. San Francisco: Believer, 2005.

Marion, Jean-Luc. *Prolegomena to Charity.* Translated by Stephen E. Lewis. Perspectives in Continental Philosophy, vol. 24. New York: Fordham University Press, 2002.

Maritain, Jacques. *The Person and the Common Good.* Translated by John J. Fitzgerald. New York: Charles Scribner's Sons, 1947.

Martin, Richard T. "Language Specificity as Pattern of Redemption in *The Thanatos Syndrome.*" *Renascence: Essays on Values in Literature* 48, no. 3 (Spring 1996): 208–23.

McCabe, Tom. "Humanity+ Aubrey de Grey Advocacy Prize." Humanity+, May 9, 2011.

———. "Parsons Conference Press Release." Humanity+, May 4, 2011.

McFadyen, Alistair I. *The Call to Personhood: A Christian Theory of the Individual in Social Relationships.* Cambridge: Cambridge University Press, 1990.

McKibben, Bill. *Enough: Staying Human in an Engineered Age.* New York: Times Books, 2003.

Meilaender, Gilbert. *Neither Beast nor God: The Dignity of the Human Person.* New York: Encounter Books, 2009.

———. "Thinking About Aging." *First Things: A Monthly Journal of Religion and Public Life,* April 1, 2011. http://www.firstthings.com/article/2011 / 03/thinking-about-aging.

Mitchell, C. Ben. "On Human Bioenhancements." Center for Bioethics and Human Dignity, Trinity International University, September 24, 2010. http://cbhd.org/content/human-bioenhancements.

Morrison, Toni. *The Bluest Eye*. New York: Plume Book, 1994.

———. *Conversations with Toni Morrison*. Edited by Danille Taylor-Guthrie. Literary Conversations Series. Jackson: University Press of Mississippi, 1994.

Moylan, Tom. *Demand the Impossible: Science Fiction and the Utopian Imagination*. New York: Methuen, 1986.

———. *Scraps of the Untainted Sky: Science Fiction, Utopia, Dystopia*. Boulder: Westview, 2000.

Murdoch, Iris, and Peter J. Conradi. *Existentialists and Mystics: Writings on Philosophy and Literature*. New York: Penguin Books, 1999.

Naam, Ramez. *More Than Human: Embracing the Promise of Biological Enhancement*. New York: Broadway Books, 2005.

"The New Science of Morality." Edge, July 20, 2010. http://edge.org/event /seminars/the-new-science-of-morality.

Noble, David F. *The Religion of Technology: The Divinity of Man and the Spirit of Invention*. New York: Knopf, 1997.

Nouwen, Henri J. M. *The Return of the Prodigal Son: A Story of Homecoming*. 1992. Reprint, New York: Image, 1994.

Nussbaum, Martha Craven. *Love's Knowledge: Essays on Philosophy and Literature*. New York: Oxford University Press, 1990.

———. *Poetic Justice: The Literary Imagination and Public Life*. Boston: Beacon, 1995.

———. "Virtue Ethics: A Misleading Category?" *Journal of Ethics: An International Philosophical Review* 3, no. 3 (1999): 163–201.

O'Connor, Flannery. *Collected Works*. Library of America. New York: Viking, 1988.

Orca, Surfdaddy, and R. U. Sirus. "Ray Kurzweil: The h+ Interview." *h+*, December 30, 2009. http://hplusmagazine.com/2009/12/30/ray-kurzweil-h -interview.

Orosan-Weine, Pamela. "*The Swan:* The Fantasy of Transformation versus the Reality of Growth." In "Reality Made Over," ed. Bernadette Wegenstein, special issue, *Configurations: A Journal of Literature, Science, and Technology* 15, no. 1 (Fall 2007): 17–32.

Parker, David. *Ethics, Theory, and the Novel*. Cambridge: Cambridge University Press, 1994.

Percy, Walker. *The Message in the Bottle: How Queer Man Is, How Queer Language Is, and What One Has to Do with the Other*. 1975. Reprint, New York: Picador USA, 2000.

———. *The Thanatos Syndrome*. New York: Farrar, Straus and Giroux, 1987.

Percy, Walker, and Patrick H. Samway. *Signposts in a Strange Land*. New York: Farrar, Straus and Giroux, 1991.

Phillips, Julie. *James Tiptree, Jr.: The Double Life of Alice B. Sheldon.* New York: St. Martin's, 2006.

Pieper, Josef. *On Hope.* Translated by Mary Frances McCarthy. San Francisco: Ignatius, 1986.

Pinker, Steven. "The Stupidity of Dignity." *New Republic,* May 28, 2008. http://www.tnr.com/article/the-stupidity-dignity.

Postman, Neil. *Technopoly: The Surrender of Culture to Technology.* New York: Knopf, 1992.

The President's Council on Bioethics. *Beyond Therapy: Biotechnology and Pursuit of Happiness.* New York: HarperCollins, 2003.

Ricoeur, Paul. *Oneself as Another.* Translated by Kathleen Blamey. Chicago: University of Chicago Press, 1992.

Robinson, Marilynne. *Absence of Mind: The Dispelling of Inwardness from the Modern Myth of the Self.* New Haven, CT: Yale University Press, 2011.

———. *The Death of Adam: Essays on Modern Thought.* Boston: Houghton Mifflin, 1998.

———. *Gilead.* New York: Farrar, Straus and Giroux, 2004.

Rosenberg, Liz. "'The Best That Earth Could Offer': 'The Birth-Mark,' a Newlywed's Story." *Studies in Short Fiction* 30, no. 2 (Spring 1993): 145–51.

Rubin, Charles. "Human Dignity and the Future of Man." In *Human Dignity and Bioethics: Essays Commissioned by the President's Council on Bioethics,* 155–72. Washington, DC: President's Council on Bioethics, 2008.

Rucker, Mary E. "Science and Art in Hawthorne's 'The Birth-Mark.'" *Nineteenth-Century Literature* 41, no. 4 (1987): 445–61.

Saunders, George. *In Persuasion Nation: Stories.* New York: Riverhead Books, 2006.

Schaub, Thomas. "An Interview with Marilynne Robinson." *Contemporary Literature* 35, no. 2 (Summer 1994): 231–51.

Scott, Kristi. "Transhumanism vs./and Posthumanism." Institute for Ethics and Emerging Technologies, July 14, 2011. http://ieet.org/index.php/IEET /more/scott20110714.

Shelley, Mary Wollstonecraft. *Frankenstein.* Everyman's Library. New York: Knopf, 1992.

Shenk, Joshua Wolf. "What Makes Us Happy?" *Atlantic* (June 2009). http:// www.theatlantic.com/magazine/archive/2009/06/what-makes-us-happy / 7439.

Silver, Lee M. *Challenging Nature: The Clash of Science and Spirituality at the New Frontiers of Life.* New York: Ecco, 2006.

———. *Remaking Eden: Cloning and Beyond in a Brave New World.* New York: Avon Books, 1997.

Singer, Peter. *Writings on an Ethical Life.* New York: Ecco, 2000.

Sklenicka, Carol. *Raymond Carver: A Writer's Life.* New York: Scribner, 2009.

Slater, Lauren. "Dr. Daedalus." In *The Best American Science Writing 2002,* edited by Matt Ridley, 1–20. New York: HarperCollins, 2002.

Spaemann, Robert. *Persons: The Difference between 'Someone' and 'Something.'* Oxford Studies in Theological Ethics. Oxford: Oxford University Press, 2006.

Stevens, Wallace. *The Collected Poems of Wallace Stevens.* New York: Vintage, 1990.

Stock, Brian. "Ethics and the Humanities: Some Lessons of Historical Experience." *New Literary History: A Journal of Theory and Interpretation* 36, no. 1 (Winter 2005): 1–17.

Stock, Gregory. *Redesigning Humans: Our Inevitable Genetic Future.* Boston: Houghton Mifflin, 2002.

Stross, Charles. *Glasshouse.* New York: Ace, 2006.

Stump, Eleonore. *Wandering in Darkness: Narrative and the Problem of Suffering.* Oxford: Oxford University Press, 2010.

Sykes, John. *Flannery O'Connor, Walker Percy, and the Aesthetic of Revelation.* Columbia: University of Missouri Press, 2007.

Taub, Gadi. "On Small, Good Things: Raymond Carver's Modest Existentialism." *Raritan: A Quarterly Review* 22, no. 2 (Fall 2002): 102–19.

Taylor, Charles. *Sources of the Self: The Making of the Modern Identity.* Cambridge, MA: Harvard University Press, 1989.

Tiptree, James, Jr. *Her Smoke Rose Up Forever.* Edited by Jeffrey D. Smith. San Francisco: Tachyon, 2004.

Tirrell, Lynne. "Storytelling and Moral Agency." *Journal of Aesthetics and Art Criticism* 48, no. 2 (Spring 1990): 115–26.

Tocqueville, Alexis de. *Democracy in America,* vol. 2. Translated by Henry Reeve. Electronic Classics Series. Hazleton: Pennsylvania State University, 2002. http://www2.hn.psu.edu/faculty/jmanis/toqueville/dem-in-america1.pdf.

Transcendent Man. Directed by Barry Ptolemy. Ptolemaic Productions, 2009.

"2008: What Have You Changed Your Mind About? Why?" Edge, January 10, 2008. http://edge.org/annual-question/what-have-you-changed-your-mind -about-why.

Vander Weele, Michael. "Marilynne Robinson's *Gilead* and the Difficult Gift of Human Exchange." *Christianity and Literature* 59, no. 2 (Winter 2010): 217–39.

Vine, W. E.. *The Expanded Vine's Expository Dictionary of New Testament Words.* Edited by John R. Kohlenberger with James A. Swanson. A special ed. Minneapolis: Bethany House, 1984.

Volf, Miroslav. *Exclusion and Embrace: A Theological Exploration of Identity, Otherness, and Reconciliation.* Nashville: Abingdon, 1996.

————. "God, Justice, and Love: The Grounds for Human Flourishing." *Books and Culture* (January/February 2009): 28–29.

von Balthasar, Hans Urs. *Love Alone Is Credible.* Translated by D. C. Schindler. San Francisco: Ignatius, 2004.

VTech. "VTech Debuts Innovative 2010 Products That 'Touch' the Future of Learning and Fun." February 11, 2010. http://www.vtechkids.com/assets /data/press/{FA989806-FB4C-496C-8F22-1A792265548D}/releases / 2011-02-13-VTech_InnoPad_Release.pdf.

Warner, Toby. "Interview with George Saunders." *Boldtype,* 2008. http:// boldtype.com/11995.html.

Waters, Brent. *From Human to Posthuman: Christian Theology and Technology in a Postmodern World.* Ashgate Science and Religion Series. Burlington, VT: Ashgate, 2006.

Weber, Brenda R., and Karen W. Tice. "Are You Finally Comfortable in Your Own Skin? The Raced and Classed Imperatives for Somatic/Spiritual Salvation in *The Swan.*" *Genders* 49 (2009).

Wegenstein, Bernadette. "Introduction." In "Reality Made Over," ed. Bernadette Wegenstein, special issue, *Configurations: A Journal of Literature, Science, and Technology* 15, no. 1 (2007): 1–8.

Weinstone, Ann. *Avatar Bodies: A Tantra for Posthumanism.* Electronic Mediations vol. 10. Minneapolis: University of Minnesota Press, 2004.

Westphal, Merold. "Levinas and the Immediacy of the Face." *Faith and Philosophy: Journal of the Society of Christian Philosophers* 10, no. 4 (1993): 486–502.

Wineapple, Brenda. "Nathaniel Hawthorne, 1804-1864: A Brief Biography." In *A Historical Guide to Nathaniel Hawthorne,* edited by Larry J. Reynolds, 26. New York: Oxford University Press, 2001.

Wong, Shelley. "Transgression as Poesis in *The Bluest Eye.*" *Callaloo: A Journal of African American and African Arts and Letters* 13, no. 3 (Summer 1990): 471–81.

Young, Simon. *Designer Evolution: A Transhumanist Manifesto.* Amherst, NY: Prometheus Books, 2006.

Zizioulas, Jean. *Being as Communion: Studies in Personhood and the Church.* Contemporary Greek Theologians, vol. 4. Crestwood, NY: St. Vladimir's Seminary Press, 1985.

Zoloth, Laurie. "I Want You: Notes Toward a Theory of Hospitality." In *The Ethics of Bioethics: Mapping the Moral Landscape,* edited by Lisa A. Eckenwiler and Felicia Cohn. Baltimore: Johns Hopkins University Press, 2007.

Index

233

Christina Bieber Lake

is professor of English at Wheaton College.

She is the author of *The Incarnational Art of Flannery O'Connor.*